This book is to be returned on
or before the date [illegible]ed below

SYMPOSIA OF THE INSTITUTE OF BIOLOGY NO. 23

Population control by social behaviour

Proceedings of a symposium held at the Royal Geographical Society, London, on 20 and 21 September 1977

Edited by

F. J. Ebling

Department of Zoology,
Sheffield University, England

and

D. M. Stoddart

Department of Zoology, King's College, London, England

INSTITUTE OF BIOLOGY
41 Queen's Gate, London SW7 5HU

First published 1978
ISBN: 0 900490 10 1
ISSN: 0537-9032

Filmset in 10 on 12 pt Compugraphic Mallard and printed in Great Britain
by A. Wheaton & Co. Ltd., Exeter

Contents

Contributors

Session 1

JOHN EBLING (Chairman) *Department of Zoology, University of Sheffield, Sheffield, England*

V. C. WYNNE-EDWARDS *Torphins, Aberdeenshire, Scotland*

JOHN KREBS *Edward Grey Institute of Field Ornithology, Oxford, England*

CHRISTOPHER PERRINS *Edward Grey Institute of Field Ornithology, Oxford, England*

J. R. FLOWERDEW *Department of Applied Biology, University of Cambridge, Cambridge, England*

Session 2

V. C. WYNNE-EDWARDS (Chairman) *Torphins, Aberdeenshire, Scotland*

ALISON JOLLY *School of Biological Sciences, University of Sussex, Falmer, Brighton, England*

J. M. DEAG *Department of Zoology, University of Edinburgh, Edinburgh, Scotland*

D. MICHAEL STODDART *Department of Zoology, King's College, London, England*

Session 3

R. D. MARTIN (Chairman) *Wellcome Laboratories of Comparative Physiology, Institute of Zoology, London, England*

J. R. CLARKE *Department of Agricultural Science, University of Oxford, Oxford, England*

J. H. MACKINTOSH *Sub-Department of Ethology, University of Birmingham Medical School, Birmingham, England*

L. R. TAYLOR *Rothamsted Experimental Station, Harpenden, England*

R. A. J. TAYLOR *Imperial College Field Station, Silwood Park, Ascot, England*

Session 4

G. E. FOGG (Chairman) *Marine Science Laboratories, Menai Bridge, Wales*

JOSEPH B. BIRDSELL *Department of Anthropology, University of California, Los Angeles, USA*

VALERIUS GEIST *Faculty of Environmental Design, University of Calgary, Calgary, Alberta, Canada*

Discussion

BARBARA BLAKE *19 Oakwood Court, London W14, England*

J. CHERFAS *Sub-Department of Animal Behaviour, Cambridge, England*

JULIET CLUTTON-BROCK *British Museum (Natural History), London, England*

J. COHEN *Department of Zoology, University of Birmingham, Birmingham, England*

J. GIPPS *Zoology Department, Royal Holloway College, Englefield Green, Surrey, England*

P. J. GREENWOOD *Biology Building, University of Sussex, Falmer, Brighton, England*

A. H. HARCOURT *Sub-Department of Animal Behaviour, Cambridge, England*

ROSEMARY E. JELLIS *Woodway, Pinner Hill, Pinner, Middlesex, England*

P. A. JEWELL *21 Hollycroft Avenue, Hampstead, London NW3, England*

C. J. KREBS *Institute of Terrestrial Ecology, Banchory, Kincardineshire, Scotland*

E. C. SAXON *Department of Anthropology, University of Durham, Durham, England*

HEIDI H. SWANSON *Department of Anatomy, University of Birmingham, Birmingham, England*

S. C. TAPPER *The Game Conservancy, Burgate Manor, Fordingbridge, Hants, England*

H. E. VAN DE VEEN *Kadijk 7, Terwolde (Gld), The Netherlands*

S. J. WALLIS *1 Victor House, Marlborough Gardens, London N20, England*

R. WRANGHAM *King's College, Cambridge, England*

A. ZAHAVI *Department of Zoology, University of Tel Aviv, Israel*

Preface

The intent of the Symposium reported in this book was to examine how far animal populations are controlled by intra-specific mechanisms, and, in particular, to debate the status of Wynne-Edwards' massive hypothesis that such control involves contests for social status, territory, or other conventional prizes rather than direct competition for food. We would like to thank all the contributors and, above all, Professor V. C. Wynne-Edwards for his opening paper and his willing participation.

Our thanks are also due to Mr D. J. B. Copp and his staff at the Institute of Biology. Lastly, we owe gratitude for the skilled help of Helen Johnson of the Institute of Biology in preparing the book, and of Elaine Bartlett, of the University of Sheffield, for transcribing the discussion and for valuable secretarial support.

JOHN EBLING
MICHAEL STODDART

Errata

The following corrections to the papers in this volume have been notified since the papers were printed.

Page 37, line 22: 'arrival' should read 'survival'.
Page 71, 7th column of Table 1, opposite *Propithecus verreauxi verreauxi:* for '0.2' read '0.02'; for '0.5' on the line below, read '0.05'.
Page 71, 7th column of Table 1, opposite *Propithecus verreauxi coquereli:* for '6.8-8.5' read '0.068-0.085'.
Page 76, 9th line from foot: for 'feeling' read 'feeding'.
Page 91, line 23: 'before invoking kin selection or' should read 'or kin selection before invoking'.
Page 92, last line: insert 'can' before 'legitimately'.
Page 94, line 23: 'chemical measures of paternity' should read 'chemical paternity exclusion'.
Page 202: Equation (2) should read:

$$N_{t+1} = N_t \exp\{R(1 - N_t/K')\}.$$

Prologue

Are animal populations controlled by intraspecific mechanisms, so that their survival is not threatened by over-exploitation of resources? The view that such homeostatic mechanisms exist 'in astonishing profusion and diversity, above all in the two great phyla of arthropods and vertebrates' was first put forward nearly twenty years ago by V. C. Wynne-Edwards. In his monumental book *Animal Dispersion in Relation to Social Behaviour*, published in 1962, Wynne-Edwards argues that although food might be the *ultimate* factor in determining animal numbers, it could not be invoked as the *proximate* agent, without disastrous consequences. A 'need for restraint in the midst of plenty must apply to all animals whose numbers are ultimately limited by food, whether they are predators in the ordinary sense of the word or not'. This restraint is provided, according to Wynne-Edwards, through social organization, which replaces the direct undisguised contest for food by competition for conventional prizes such as social status, territory, and the like, which in turn determine each individual's right to survive and reproduce.

Animal Dispersion in Relation to Social Behaviour consists of a comprehensive and elaborate documentation of all the data relevant to the hypothesis. Having first reviewed at length the visual, sound, electric, scent, and tactile signals which are used for social integration throughout the animal kingdom, Wynne-Edwards proceeds to discuss the role of threat in establishing social hierarchies, whose 'function (not hitherto defined) is always to distinguish the "haves" from the "have-nots" whenever the population density becomes excessive'. This conclusion was reached before the publication of Konrad Lorenz's book *On Aggression;* Wynne-Edwards makes no reference to Lorenz (apart from a single paper written in 1931), and quotes only Hingston's 'thought-provoking book *The Meaning of Animal Colour and Adornment*' published in 1933. Wynne-Edwards then reviews territorial phenomena, noting that they are especially common in birds, but occur also in fish, amphibia and reptiles, mammals (including primitive man), and arthropods. A mosaic of territories can prevent the population density from rising higher once the habitat is completely occupied, by keeping surplus individuals out. For homeostatic purposes territories would need to be adjustable in size, in inverse proportion to the amount of food they can yield; and this prediction has been tested experimentally and confirmed for the red grouse (see pages 12–17).

The renewal of food resources is often seasonal and subject to irregularities which alter the carrying capacity of the habitat. In order to get the best homeostatic adjustment to these changes the consumers need a feedback of information of two kinds. One is an appraisal of current food resources, which individuals may sometimes derive simply from their own changing state of nutrition, and the other is an indication of the number of mouths to be fed, and is experienced through social, density-dependent pressures from fellow consumers. For populations not dispersed in territories where they have neighbours to keep up the pressure, but living a roving life instead, alone or in groups, special *ad hoc* assemblies with more or less dramatic mass displays, for instance at a communal roost, seem to be necessary to provide the social feedback. These, and other displays that appear to fulfil the same ends, Wynne-Edwards calls epideictic displays.

Wynne-Edwards points out that exactly the same social mechanisms that govern the individual's right to feed and live in the habitat, namely property tenure and personal hierarchies, can also control his admission to breeding status, and thus regulate recruitment into the adult class. The existence of this dual control of the two basic social rights, to feed and to reproduce, adds much weight to the hypothesis that the primary function of such social adaptations is homeostatic.

In illustrating these effects he cites experimental evidence. For example when guppies (*Lebistes reticulatus*) were put into aquaria with excess food, they reproduced only to reach a ceiling biomass of 32 grams per 17 litres of water (Silliman and Gutsell, 1958). (There is evidently a maximum crowding threshold acting here in place of a variable food-related one.) Similar results were obtained with rodents, in which homeostatic restraints could be produced either by overcrowding *per se* in the presence of excess of food, or by limitation of food alone.

Changes in reproductive output, for example in pregnancy rates, intra-uterine resorption, and litter-size, could be mediated by neuroendocrine mechanisms according to several authors, including Clarke (1953) who is a contributor to the present symposium. A relationship between reproductive output and population density was also evident from field studies of birds. For example Kluyver (1951) showed that over a period of several years both clutch size and the percentage of pairs producing second broods in the great tit (*Parus major*) were negatively correlated with density.

Such homeostatic mechanisms arise in evolution by what Wynne-Edwards calls group selection. 'Evolution can be ascribed, therefore, to what is here termed group-selection—still an intraspecific process and, for everything concerning population dynamics, much more important than selection at the individual level. The latter is concerned with the physiology and attainments of the individual as such, the former with the viability and survival of the stock or race as a whole. When the two conflict, as they do when the short-term advantage of the individual undermines the future safety of the race, group selection is bound to win, because the race will suffer and decline, and be supplanted by another in which antisocial advancement of the individual is more rigidly inhibited.'

The idea of group selection was clearly central to Wynne-Edwards' original argument. While agreeing that it was not by any means a new concept, he commented that it had never been accorded the general recognition that its importance deserved, and he returned to it many times throughout his book. Nowhere, however, did he present any rigorous biological model. It is interesting to note his statement that a hypothesis due to Carr-Saunders (1922), namely that all primitive peoples limit their numbers to an optimum population density by such cultural practices as abstention from intercourse, abortion, and infanticide, depends on 'social evolution through group selection'. Wynne-Edwards thus made no distinction between the quite reasonable view that cultural systems may be selected by the advantages they confer on the social groups which employ them and the more contentious thesis that heritable behavioural traits which reduce individual fitness in favour of the species as a whole may evolve by natural selection.

The attraction of Wynne-Edwards' hypothesis is undeniable. That over-exploitation of resources in the short term must be disastrous to the ultimate survival of animal and human societies alike seems axiomatic, and no intellectual effort is required to extend the concept of homeostasis, already familiar as applied to the internal environment of the organism, into the environment of the society. At the same time, the hypothesis supplies a functional explanation for a range of behavioural phenomena in animals, including particularly that known as ritualized aggression and given wide publicity by Konrad Lorenz (1966). With growing apprehension about the explosive growth of human populations it seemed, moreover, to point to a human moral, notwithstanding that proximate restraints on reproduction were invented by man long before the concept was considered in relation to animals. Indeed, one of Darwin's arguments for a 'struggle for existence' was that animal reproduction was not subject to any 'prudential restraint by marriage' which Malthus had previously recognized as a human attribute.

For all these reasons, the hypothesis that animal populations were controlled by intraspecific behavioural interactions became readily absorbed into popular ethology, in which a growing interest was fostered by such books as *The Territorial Imperative* and *The Social Contract* by Robert Ardrey (1967, 1970). Nevertheless, there were immediate criticisms, both of the hypothesis itself and of the idea that 'altruistic' traits could be produced by 'group selection'.

Lack (1966) was among the earliest critics. He rejected the view that food was a limiting factor, stating that 'Wynne-Edwards was also wrong to consider, and was mistaken in quoting me as considering, that the vast majority of animals are limited by food. Food admittedly sets an upper limit to all animal populations, but as already pointed out, this limit seems rarely if ever to be reached under natural conditions.' He similarly rejected a role for group selection, remarking: 'It is generally extremely hard to see how "altruistic" behaviour, which decreases the chance of survival of the individuals concerned but increases the chance of survival of others in its group, could have a selective advantage.'

Parental behaviour, as Lack recognized, is an obvious

exception, because natural selection will favour those hereditary types which leave most relatives. So is altruistic behaviour in favour of several close relatives, since these will each have a greater than average chance of sharing the same gene, and thus the result will be an increase in the 'total fitness' of the kin group, to use a concept invented by Hamilton (1964). 'Kin selection', by which such altruism could arise, is no more than an extension of orthodox natural selection. The minimum requirement for a suicidal altruistic gene to be successful is that it should save more than two children or parents, four uncles, aunts, nephews, nieces, or grandchildren, or eight first cousins.

Recently, indeed since our Symposium was planned, Dawkins (1976) has carried the argument further in a lucid, polemical attack on group selection. He and his fellow critics do not deny that restraints to fertility occur in overcrowding, but they argue that these arise not as mechanisms to limit population in the interests of the species, but to further the chance of survival of the genes carried by the individual.

Dawkins agrees with Lack in believing that animals will tend to have the optimum number of children from their own selfish point of view. If they bear either too few or too many, they will end up rearing fewer, and thus propagating fewer of their genes than if they had hit on the right number. The optimum is likely to be smaller in a year of overcrowding than in a year of sparse population, since overcrowding is likely to foreshadow famine. There will thus be a selective advantage in favour of the female who can respond to overcrowding by reducing her own birth rate. Dawkins similarly uses the 'selfish gene' argument to explain 'epideictic' displays. He suggests, for example, that the roosting of starlings, in which birds are exposed to each other's calls, might enable each female bird to 'predict' what is going to be the optimum clutch size for her, as a selfish individual, in the next breeding season. He goes on to point out that it will also be to the advantage of each selfish individual to make as much noise in the winter roost as possible, in order to fool other starlings into reducing their clutch sizes disproportionately—what J. R. Krebs has called the 'Beau Geste Effect'.

It appears as if the selfish gene explanation can be applied to almost every eventuality, though not always with equal facility. Consider, for example, the fact that pregnant female mice are induced to abort by the odour of a strange male, an effect first identified by Bruce (1965). The phenomenon, though not originally mentioned by Wynne-Edwards, was quoted by Ardrey (1970) as a mechanism for population control in conditions of overcrowding. Such a functional explanation is dismissed by Dawkins, who proposes that, by inducing the pregnant female to abort, the strange male destroys his potential step-children and can then utilize the female to further survival of his own genes. The situation is analogous to that of male lions who, when newly arrived in a pride, usually murder all existing cubs, presumably because they are not their own offspring.

But is it? Dawkins' explanation seems singularly male-orientated. The equation to be solved by the population geneticist may be more complicated for the mouse than for the lion. If the male mouse has evolved the odour, it is the female which has evolved the response by which her own fecundity is reduced. Why do not her own selfish genes resist such a development? One possible answer is that the sacrifice of a modest amount of reproductive time will be more than counterbalanced by the future advantage of associating some of her genes with a more odoriferous, and hence more abortigenic, male line.

It may be noted that for such an evolutionary development to occur, the successful males need have no superiority other than their potent odour. While we have no evidence about the operation of the Bruce effect in wild populations, it could well be density dependent, since the greater the overcrowding the more probable it is that any pregnant female would be assaulted by the odour of a strange male. In overcrowded circumstances the gene of the strange male would not only secure its own priority but, by reducing competition for resources, further enhance its chances of survival. In short, the very selfishness of the gene would act to reduce overall reproductive capacity and thus favour the species by promoting survival of at least some members of the population. In essence this is exactly consonant with Ardrey's functional explanation!

Such is the background to our Symposium. Our objectives were to assess the current evidence for intrinsic population control and the possible mechanisms by which it could be brought about. Populations are not necessarily isolated and we extended the scope by inclusion of the elegant papers on mobility by Flowerdew and by the Taylors, father and son. But believing it would be more important to examine the postulated phenomena than to discuss evolutionary theory, we did not invite any papers on population genetics, though at the same time the topic was not excluded from discussion.

In his opening paper, Wynne-Edwards himself assesses the evidence which has accumulated in the last twenty years. In particular, he reviews his project on the red grouse, concluding (p. 16) that the results 'leave no doubt that the grouse control their own population density, and that they vary the average size of the territories they take, from place to place and year to year, in conformity with the state of their food resource'. At the same time, he notably admits the failure to invent an adequate mechanism for group selection with the words (p. 19) 'The general concensus of theoretical biologists at present is that credible models cannot be devised, by which the slow march of group selection could overtake the much faster spread of selfish genes that bring gains in individual fitness. I therefore accept their opinion.'

The Taylors' paper (p. 181) attempts to fill the lacuna by using a spatial mechanism to connect Wynne-Edwards' evidence for density-dependent behaviour with intrinsic population control. They see homeostasis as the outcome of competing ecological pressures on the individual. These pressures are reflected in population density, measured by competition for mates, in epideictic display or, in more primitive organisms, by simple contact or competition for resources. Each individual responds to population density by moving, with more or less success, in response to the dilemma presented by, on the one hand, the advantages of lower competition but poorer resources where density is low and, on the other hand, better resources but more competition where the density is high. The individual's success in making the choice is judged by its children's survival to reproduce in their new home. The resulting density-dependent movements within and between populations make populations mobile in response to environmental change and, when simulated, they stabilize below environmental limits without altruistic restraint.

Krebs and Perrins summarize the fieldwork carried out over thirty years in both England and Holland on the great tit. The evidence appears to support Wynne-Edwards: territoriality in the spring and aggressive encounters between young birds in the late summer have major effects on population density. However, the authors state (p. 44) that these effects were incidental and that territoriality evolved as a mechanism of individual competition for resources.

Flowerdew adds another dimension to the problem in his consideration of wood mice, by pointing out that transients as well as residents are concerned in their population dynamics. Populations vary little from winter to winter, but quite considerably from summer to summer. The winter numbers appear to be adjusted to provide a viable breeding population for the rest of the year, without completely 'overeating the winter food supply' (p. 63). It seems that this is achieved by a density-dependent process, involving social interactions, which is initiated in summer.

The next two papers deal with primates, about which virtually no information relevant to Wynne-Edwards' hypothesis was available at the time *Animal Dispersion in Relation to Social Behaviour* was first published. Although primatologists now take for granted that hierarchy and territory regulate animals' access to key resources, this situation has changed little. Among Malagasy lemurs, according to Jolly (p. 72) the white sifaka, *Propithecus verreauxi*, lives in territorial groups, maintained all the year round, though with some intergroup migration and some overlapping, and its population is fairly stable. *Lemur catta* maintains an even more stable number, though its territorial ranges vary and at some periods almost wholly overlap. The most that can be concluded is that the two species are ripe for further study. Deag, in a provocative paper which reviews a number of behavioural interactions in baboons and macaques, not only refuses to make any generalizations about their functional significance, but warns others of the danger of so doing. Three of the remaining papers deal with possible mechanisms of population homeostasis. Mackintosh describes the experimental analysis of overcrowding in rodents and discusses some of the physiological changes, for example, the increased adrenocortical and decreased gonadal activity in subordinate mice after aggressive encounters, a theme developed in detail by Clarke. Stoddart enlarges upon the role of odours in mammalian communication, especially in the establishment and labelling of social dominance, the demarcation of territory, and the transmission of danger and warning signals.

Finally Birdsell and Geist turn attention to Man, the one providing factual ecological information about a contemporary primitive people, and the other a speculative essay on the possible adaptive life styles of our Palaeolithic ancestors. Birdsell shows that Australian Aborigines provide a beautiful model in which population densities are highly correlated with such environmental variables as rainfall. Optimal numbers have been maintained by behavioural mechanisms operating at the levels of the tribe, local group, and family, which include totemism, initiation rites for young men, kinship networks, the judicial system, and preferential female infanticide. Such findings amply support the hypothesis of Carr-Saunders, but this must be seen as analogous to, not homologous with, that of Wynne-Edwards. For the rest we must leave readers to assess for themselves the current status of Wynne-Edwards' theory from the detailed papers which follow and the liveliness of the printed discussion.

JOHN EBLING

References

Ardrey, R. (1967) *The Territorial Imperative*. London: Collins.

Ardrey, R. (1970) *The Social Contract*. London: Collins.

Bruce, H. M. (1965) Olfactory pheromones and reproduction in mice. Proc. IInd Int. Cong. Endocr. Amsterdam, *Excerpta Medica*. pp. 193–197.

Carr-Saunders, A. M. (1922) *The Population Problem: a Study in Human Evolution*. Oxford.

Clarke, J. R. (1953) The effect of fighting on the adrenals, thymus and spleen of the vole (*Microtus agrestis*). *Journal of Endocrinology*, **9**, 114–126.

Darwin, C. R. (1859) *On the Origin of Species*. London: John Murray.

Dawkins, R. (1976) *The Selfish Gene*. Oxford: Oxford University Press.

Hamilton, W. D. (1964) The genetical theory of social behaviour (I and II). *Journal of Theoretical Biology*, **7**, 1–16; 17–32.

Kluyver, H. N. (1951) The population ecology of the great tit, *Parus m. major* L. *Ardea*, **39**, 1–135.

Lack, D. (1966) *Population Studies of Birds*. Oxford: Clarendon Press.

Lorenz, K. Z. (1966) *On Aggression*. London: Methuen.

Silliman, R. P. and Gutsell, J. S. (1958) Experimental exploitation of fish populations. *U.S. Fish and Wildlife Service Fishery Bulletin*, **58**, 214–252.

Wynne-Edwards, V. C. (1962) *Animal Dispersion in Relation to Social Behaviour*. Edinburgh and London: Oliver and Boyd.

Intrinsic population control: an introduction

V. C. WYNNE-EDWARDS

Torphins, Aberdeenshire, Scotland

Introduction

I appear to have been the first person to suggest that there was a general connection between population control in animals and their social behaviour (Wynne-Edwards, 1959). It therefore falls to me to introduce the subject here. Twenty years ago it was just an idea, a hypothesis. Research then begun, however, has revealed that at least in one species of bird, the red grouse (*Lagopus lagopus scoticus*), the hypothesis is experimentally verifiable, and population density can be shown to be determined by the social behaviour of the birds themselves.

The behaviour that brings this about is the taking up of territories by the cock birds in the autumn, in a contest in which many birds are unsuccessful. From then on, for the next nine months, the successful ones own and occupy the habitat in an almost complete mosaic. The unsuccessful on the other hand are reduced to the status of outcasts, liable wherever they go to be chased off the heather on which they feed. As winter advances their condition deteriorates and they fall victims to predators and starvation. Outcast hens meet with the same fate as the outcast cocks, and only rarely does a surplus bird remain alive when spring arrives. I shall have more to say about this later.

There were others before me who suggested that territoriality could impose a ceiling on population density, and thus account for the relative stability of bird numbers which we can generally observe from year to year. This was postulated by the Irish naturalist C. B. Moffat in fifteen pages of print as early as 1903. What the earlier writers did not do, however, was to recognize that territoriality is essentially a social phenomenon. Consequently they did not take the succeeding step and notice that there is another common component of social behaviour which achieves the same result. This is the social hierarchy, the ranking of individual members of a social group into a pecking order. It has the same ecological effect of excluding subordinate individuals, when necessary, from safety, food, or reproduction, and sometimes of driving them out of the group and the habitat.

Darwin's checks

When animals live under more or less natural conditions, their numbers, and especially those of the larger animals, change comparatively little from year to year. Even the populations that fluctuate appreciably tend to do so about a mean level. There are of course exceptions, but the rule of stability is general enough to suggest that population density is not a random variable, and consequently demands an explanation. Darwin addressed himself to this problem in *The Origin of Species*, noting on one side of the population balance-sheet a general and sometimes prolific power of increase through reproduction, and on the other side a range of lethal agents that appeared to be holding the populations in check. Darwin grouped these checks into four categories: first, the food supply, which must always dictate an absolute upper limit; second, predation or the serving as food for others; third, the climate, meaning the fluctuations of the physical environment; and fourth, epidemic disease. With his usual prescience he noted that epidemics tend especially to appear when numbers are high, meaning, in present-day terms, that they have an element of density dependence. But he was careful to acknowledge that 'the causes which check the natural tendency of each species to increase are most obscure', implying that these checks are not necessarily the true or the whole explanation of population stability.

Density dependence is a necessary attribute of any population control factor. It implies that the further the population density gets from the norm or optimum level, the more powerfully do the factors operate to restore it to that norm. As Lack pointed out (1954, p. 19), in theory either the output from reproduction or the losses due to mortality could be density dependent: that is to say that faster reproduction could be induced by lower densities, or heavier mortality by higher densities.

My thoughts on population regulation received a powerful

stimulus from reading Lack's book in 1954. I had already for long been familiar with the overfishing problem, first in relation to the commercial fisheries of the North Sea, and the whaling and fur-seal industries, and later to sport fishing, wildfowling, and fur-trapping in Canada. I was aware that as a rule, where man had found a profitable living as a predator on populations of wild animals, he had tended to overexploit them, and that the accepted way to prevent this happening was to impose restrictions, such as close times and bag limits, quotas on the numbers of exploiters, or controls on the lethal capacity of the equipment they were allowed to use. Where it was possible to introduce it, the ideal method was to divide the habitat into territories each belonging to one person, who could expect to pass his territory on to his heirs. Only then would prudential management become paramount and productive capacity be jealously conserved. The exploitation of grazing land by herbivores is just as prone to abuse if the stock density is allowed to rise too high; hence the traditional assignment of farmland to hereditary occupiers.

The idea that a given habitat has a finite carrying capacity, in terms of the numbers of particular consumers that its productivity can sustain, is familiar to ecologists. It should be said in passing that carrying capacity is related strictly to what the prey or plant food resources can stand, and these are not of course the only resources that animals need in a habitat. Where one of the other necessities of life is limiting or absent, such as breeding or refuge sites, the potential carrying capacity of the food supply will not be reached.

We know that man ultimately discovers whether he is exceeding the carrying capacity of the living resource he is exploiting, only by trial and error. There is usually no natural sign, no 'warning light', to show him when the maximum sustainable yield has been reached. Because the 'noise' in the environment causes considerable changes in crop yields from year to year, it may take a long time before the results of overexploitation are recognized; and if depletion is still not obvious within a lifetime of human experience, he may inherit the belief that poorer harvests are a fact of nature, and never realize the underlying cause. Thus neolithic pastoral tribes in arid lands probably watched the desert fringe advance for many generations without realizing that they had a remedy against it in their own hands.

In general terms, animal predators and herbivores are faced with the same problem as man. I have several times illustrated this by reference to the beaver in North America (*Castor canadensis*). It is of course a herbivorous rodent which crops whole trees, especially for winter food. Its staple food is generally the aspen (*Populus tremuloides*), which it fells in advance in the late summer and autumn, and floats to the beaver dam. There the logs and brushwood are stock-piled in the pond, with the base of the heap resting on the bottom. When the pond is frozen and snow-covered in winter, as it may be for up to eight months of the year in northern Canada, the beavers have dive-ways leading from their lodge, out under the ice, through which they can swim to bring back the boughs, whose bark and buds provide their food, to be eaten in the lodge.

Considered as a food resource, whole trees have a long regeneration time from one felling to the next. After they have been cut down by beavers aspens renew their growth relatively quickly compared with some trees by sending up suckers; but measurements made in Manitoba (Green, 1936) suggest that the average tree felled is 11–12 cm in diameter, and is probably twenty years old or more. That means that beavers must be harvesting the tree crop on roughly a twenty-year minimum rotation. The same figure can be turned the other way up by saying that the maximum sustainable yield of aspens is 1/20th of the available trees per year. In an average year, therefore, 19 out of 20 trees must be left standing if the habitat's productivity is to be conserved. Beavers are popularly considered to be remarkable mammals, from their habits of cutting down trees, constructing dams and canals for transport, and laying up winter stores in advance, but it is still more surprising to realize that, though they normally have twenty years food supply around them just waiting to be felled, they are inhibited from increasing their numbers in order to take advantage of it. If they did, the resource would progressively dwindle until their habitat was rendered uninhabitable.

In the Mackenzie delta, a huge 100-mile plain of muskeg, ponds, and weaving channels, with a roughly 50-50 balance in area between land and water, beaver colonies are now rather uniformly scattered about a mile apart (Aleksiuk, 1968). The home-range actually being used by a colony is only about a quarter of what this spacing suggests, for between the colonies there are wide buffer zones, utilized only by a small floating reserve of two-year-olds, and comprising the remaining three-quarters of the delta habitat. In the course of time the colonies no doubt shift their home-ranges somewhat. Beavers do not need to build dams here because there are channels and ponds already. Living so widely spaced out may be an additional part of their adaptation for limiting their numbers in this, their northernmost habitat, which is underlain by permafrost, and where the larger trees grow only on the channel banks, with their roots bathed by river water still carrying in summer a residue of the warmth it absorbed a month or so before, on the hot prairies in Alberta 2000 miles to the south.

The ratio of annual crop to standing stock varies of course very much among the countless food species utilized by different animals. If the food takes years to attain its harvestable size, like aspen trees for beavers or herring for dolphins, the sustainable yield will be rather a small fraction of the total biomass; whereas fast-maturing and prolific insects like aphids can be consumed, at least by small birds, virtually *ad libitum*, so long as enough aphid eggs remain to overwinter and start up the next year's crop. Even where the whole stock of food can safely be eaten, it may have to be eked out to last the full period for which the consumers have to be dependent on it. Thus limiting the population density of consumers in order to assure that its members will get enough to eat, is a common ecological necessity for animals like vertebrates. Only a minority of them can retire if necessary into suspended animation, or keep pace with a boom-and-bust economy.

It was with these trophic complexities in mind that I put forward the hypothesis that many and perhaps most animals must be adapted to preserve a prudential relationship between their own population density and the current food supply, and that they make use of social competition as an artificial regulator of food demand, in a manner I am shortly going to describe.

Perhaps I have already succeeded in making it clear that for some animals it is necessary to hold back on food consumption, at least for some of the time. But of course no such restraint would be necessary if their populations were being held down by any of Darwin's other checks, namely predation, climate, or disease. Sufficient mortality from these factors could prevent them from ever reaching densities at which they could overtax the food supply. Before going further therefore it is important to examine the efficacy of these checks as population regulators.

Predation is certainly capable of holding prey populations in check for long periods, as witness the stocks of food-fishes being commercially exploited by man in the North Sea. Predation is partly density dependent, in that predators are generally known to concentrate their attention on prey that is abundant, and to leave it alone when it is scarce. The predator–prey relationship is an inherently unstable one, however. Predators linked to one staple prey, as for instance herring fishermen to herrings, are in danger of overtaxing or eventually exterminating their prey altogether if they persist in stepping up their catching effort; and what they can do to diminish one prey species they can do successively to another and another. In other words, predation is itself part of the food-web, and predators like other consumers have had to develop prudential restraints in order to manage and conserve their own food supplies. By definition, the top predators have no predators themselves, and predation can never serve them, therefore, as a check on population growth.

Climate, Darwin's next check, which we can widen to cover the whole range of variability in the physical environment, can also limit population growth. Every widespread species somewhere meets the limits of its geographic range, beyond which it may advance only when favourable changes allow, and from which it will be driven back or exterminated if harsher conditions ensue. This too is a notoriously unstable relationship, especially characteristic of the polar fringes and the desert edge; its density dependence is nil. There are species both of animals and plants that have nevertheless become adapted to opportunism, and survive as perpetual colonists in these precarious biotopes. Of the animals at least some, including the migratory locusts and lemmings, nevertheless exhibit remarkable adaptations for self-induced population change. When times are good both of these can swamp all inroads that their predators may make on population growth, and send out large suicide squads, seeking new habitats to colonize. It is their recipe for survival in a catastrophic environment. The majority of animals and plants, on the other hand, live in safer, stabler climates than these, where catastrophes are fewer, smaller, and for long periods totally ineffective as checks to population growth.

There remains communicable disease, again a factor with some semblance of density dependence. But animals and plants, in acting as hosts to disease organisms, show a remarkable ability to develop a tolerance to the invaders' presence; and because pathogens are such powerful selective agents, a state of immunity is often reached in the survivors after relatively few generations. If immunity is not achieved the affected host population succumbs, and the relationship remains so unstable that either the host or the pathogen dies out. Infective disease can rarely if ever, therefore, provide a permanent population regulator.

Self-regulation or homeostasis

If none of these extrinsic checks is reliable as a population regulator, what is the alternative? It could be to develop intrinsic mechanisms to provide a homeostatic system of population adjustment and control. The essence of this hypothesis is already familiar to many participants in the symposium, and space does not allow me to go over the ground in great detail here. A fuller summary of it can be found in Chapter 1 of *Animal Dispersion in Relation to Social Behaviour*, or in the body of the book itself (Wynne-Edwards, 1962).

In forming a mental picture of such a homeostatic system it is perhaps advisable to think of vertebrate populations as providing typical examples, although the concept does lend itself to numerous invertebrate species as well. The system needs to have two essential elements. First, the animals require an input of information, a data base, to enable them to assess the state of their food supply; and secondly they require response mechanisms that will alter their population density to match their food prospects and so keep the population within the carrying capacity of the food resources. Both processes would be largely or completely automatic.

The monitoring data base would itself appear to require an integration of two components. The first of these is the amount of food in prospect at the moment, and the second is the number of individuals present and wanting to consume it. The amount of food divided by the number of consumers gives an index of how long it will last, and whether its carrying capacity is sufficient to meet the demand. Food abundance at the moment is reflected inwardly in the nutritional condition of each individual (in its body weight, fat deposits, etc.), and outwardly in the visible food supply. The number of consumers present is indicated by the level of intraspecific competition which each of them finds in their quest for social rights and rewards; and one of these rewards is the chance to fill one's belly without being moved on by a dominant rival. In many species there appear to be also *ad hoc* demonstrations of population size, which I have called epideictic displays, such as the flight manoeuvres and the mass chorus of starlings at a roost, the synchronized dawn and dusk singing of many birds, frogs and insects like cicadas, the mass displays of hibernating monarch butterflies and aestivating bogong moths, and perhaps some of the aerial dances of midges and gnats (Wynne-Edwards, 1962).

The response mechanisms for producing population change are again different, according to whether they push the density down or push it up. There is endless variety of detail in the working of each. Surplus population is generally removed as a result of aggressive competition, for example by subdividing the habitat into individually or family-owned territories too few in number to accommodate all the contestants, and thus squeezing the surplus individuals out and driving them away. As we shall see later, at times when the food supply is poorer, aggression tends to become stronger and territories larger, forcing the population density lower. In gregarious species which do not occupy individual territories, subordinates may be displaced from the best feeding areas or otherwise harassed, so that even if they are not forced out they may still lose condition, be more easily taken by predators, or die from malnutrition or disease.

The most important way of adding to the population is of course by reproduction; but outside the breeding season it is sometimes supplemented by immigration, if the population density is low and aggression consequently relaxed. Incomers from elsewhere, or immature individuals, are then allowed to enter and augment the existing population.

The fundamental role of reproduction among life processes, and its importance to the fitness and fulfilment of the individual, are reflected in the striking fact that the individual's admission to the breeding caste is governed by the same social mechanisms that govern his access to food. Reproduction is a right and feeding is a right, and each is won or lost through competition for the same conventional prizes of property and personal status. Feeding and breeding can both be denied to individuals that are surplus to the acceptable quota, a quota *ex hypothesi* related to the expected carrying capacity of the habitat. In many animal species, failure to win a territory, or a high enough place in the local hierarchy, has profound physiological consequences in the neurendocrine system, and it frequently inhibits the maturation of the reproductive organs.

The hypothesis goes on to point out the key part that social behaviour appears to play in population homeostasis, and to suggest that it was in this role that sociality first arose in evolution. It seems to me most likely to have developed initially in relation to food, because of the necessity of imposing prudential restraints on food consumption while the resources are still plentiful. When food is plentiful for all, obviously it cannot be the object of fierce competition; it was therefore necessary to concoct a substitute prize instead, which could be competed for at a much earlier stage. Initially no doubt this prize was what Moffat called 'a parcel of ground', that is, a feeding territory. Provided the individual was adapted to claim a territory of sufficient size, the population density could be set at any level that happened to be appropriate to the current carrying capacity of the food supply. Such a territory appears to be essentially a substitute goal, in place of the food it contains; being an artificial device, it can be competed for only under artificial or conventional systems of behaviour and values, in which all the rivals are all bound by the same artificial code. It is this, in my view, which gives rise to the social phenomenon; sociality provides the conventional world in which individuals compete according to imposed rules.

In subsequent evolution, the goals of conventional competition have often become more artificial. The property goal has been reduced in many instances to a token holding, no longer directly related to food, but nevertheless still conferring on the individual the right to feed in communally held feeding grounds. Examples are the nest territories held by colonially breeding birds, or the personal nesting or display sites used by many kinds of birds, mammals, reptiles, and fish. Property qualifications have even disappeared altogether in some species, and been replaced by a purely abstract qualification of personal status, which confers rights on individuals standing sufficiently high in a hierarchy of competitors. Thus in most gregarious animals that habitually live in flocks or schools, group membership and its inherent rights depend solely on personal acceptance by more dominant individuals.

This hypothesis allows one to define a society very simply indeed, as an organization of individuals capable of providing conventional competition among its members. Though to some people it may come as a surprise to find such emphasis laid on the competitive nature of society, not much reflection is required to see that the cap fits. Do we not find, in our own social world, the leaders at the top, the winning teams, the house-proud, the status seekers, the covetors of pay-differentials, of rank, titles, and honours, which reveal the pervasiveness of sophisticated, and that is to say social, competition in our lives?

One of the assets of using social competition as a population regulator is that it is automatically and completely density dependent. The more the competitors, the greater the rivalry. It is thus able to solicit a response exactly commensurate with the need. Population homeostasis is a more or less continuous process in which the population's demand for food is adjusted to a fluctuating supply. In some very short-lived, quickly reproducing microbes and invertebrates, if there are any controls being exercised at all they are concerned solely with reproduction or recruitment. And in any very stable environment it is theoretically possible to stabilize the population density solely by making a constant annual recruitment to replace a constant natural mortality. Under these circumstances, the mortality would be the independent variable and recruitment the dependent variable in the homeostatic equation.

Such a balance can in fact be approximated to under experimental conditions with populations of certain laboratory animals, such as guppies (*Lebistes reticulatus*). But even with fish in a tank, or fruit-flies in a culture chamber, the equation is not quite as simple as this. With guppies, after a population ceiling has been reached in the experimental aquarium, the fish go on reproducing as before, but they cannibalize the newborn young, trimming the numbers down to the few that are required to compensate for the deaths of senile fish. That is to say, the fish themselves contribute to both sides of the equation, to natality and to mortality as well. If supernumerary adults are released into the experimental tank, cannibalism will get rid of a surplus of adults too and bring the population back to the previous ceiling level. With *Drosophila*, the adult flies emerging from pupation exhibit mutual aggression. An experiment showing the effects of this was made using culture cages, each in the form of a lucite box with a series of 20 portholes, into which vials containing culture medium could be fitted. By renewing the vials in rotation the *Drosophila* population could be kept going indefinitely. The unique feature of the cages was that two of the tubes were always left empty and contained no food. The numerous adult flies competed to gain access to the medium in the other vials, and the experimenter observed that more than half of all the adults soon found their way into the empty tubes, became moribund, and died there. These so-called 'death sites' actually constituted only 1/60th of the floor space of the cages. Various tests suggested that the dying flies were refugees from the territoriality and social dominance exercised by more aggressive individuals in the food-containing tubes; and that, on a miniature scale, the behaviour of the refugees corresponded to the emigration by which, in the outside world, persecuted individuals normally escape from the arena in which competition occurs (Milkman, 1975).

The members of a population can change their own population density by four different kinds of action. Two of these normally or sometimes take place *in situ*: these are reproduction, and mortality which is socially contrived. The other two involve removing individuals to or from another place, by emigration or immigration. In each set, one action leads to an increase in density and the other to a reduction. We can expand the homeostatic model equation to take account of these four processes. They can be used alone or in any suitable combination to restore a balance that has been upset by uncontrollable external forces. These forces may for example have changed the rate of mortality or altered the carrying capacity of the habitat, so that population compensations need to be made. Essentially there are five terms in the equation, two contributing to the income side and three to the loss side of the balance, as follows:

$$\begin{array}{c}\text{recruitment}\\ \text{from}\\ \text{reproduction}\end{array} + \text{immigration} = \begin{array}{c}\text{uncontrollable}\\ \text{mortality}\end{array} + \begin{array}{c}\text{socially}\\ \text{contrived}\\ \text{mortality}\end{array} + \text{emigration}$$

Of the five, only one is an independent variable, namely uncontrollable mortality; if this function, or if resource levels, change so that a new population level needs to be sought, the animals' own homeostatic adaptations supply the means of reaching a new balance, with a density higher or lower or at the same level as before. But time must be allowed for the machinery to work, and it can easily take a year, and occasionally much more, if there has been particularly devastating mortality.

Population adjustments are generally made at particular seasons of the year, especially among vertebrates and in seasonal climates. This applies particularly to recruitment from reproduction, to the eviction of surplus individuals, and to major migratory movements such as those of many birds. It would appear to be particularly significant, therefore, to find that just before or during these changes, social competition and epideictic displays tend to build up to their highest intensity.

Testing the hypothesis: the red grouse research

While I was writing the book *Animal Dispersion* it naturally appeared to be an urgent matter to put the hypothesis to the test, so far as that was practicable. In 1956 an opportunity came to obtain funds for an applied research project on the red grouse (*Lagopus l. scoticus*), to try to discover why it had become so much less numerous on the Scottish moors than it had been twenty or thirty years before, and what if anything could be done to reverse the decline. The key to this problem would obviously lie in finding an answer to the more fundamental question of what controls the population density of the red grouse.

A small but strong research team was quickly assembled to look into it. Now, 21 years later, the Grouse Research Unit is still in existence, and many of the attainable answers have been found; but it was no easy task, and seven years elapsed before the first major publication appeared (Jenkins *et al.*, 1963).

Grouse are almost wholly vegetarian birds, and this particular subspecies depends for nine-tenths of its food on the dominant plant in its habitat, *Calluna vulgaris*, the ling heather. It is a low heathy shrub covering 40–75 per cent of the ground on good grouse moors, and the birds eat its leaves, buds, flowers, and seeds. Grouse population densities in north-eastern Scotland commonly lie between 50 and 250 birds per km^2, that is, between one bird to five acres and one bird to every acre. For such low densities, food appears at first sight enormously superabundant. The research team were highly sceptical of the idea that density would prove to be governed by the food potential of the habitat. But gradually events began to suggest that this might indeed be true. A biochemist, Robert Moss, joined the team, and he showed that though the birds were never short of energy-giving carbohydrates, with proteins it was a different matter. Grouse are in fact very selective in what they eat. He watched them feeding and filling their crops, and then tried to imitate them by picking off similar tips of young growth from the same plants, and he found that the nitrogen content of the undigested food in the birds' crops was 10 per cent higher than it was in his own sample (Moss, 1967). In addition, the nitrogen content of the heather growing on acid granite soils was found to be lower than it is on more neutral soils derived from diorites and limestones, although the plants look just the same to the eye. Grouse populations are often several times as dense on the latter as on the former.

Even on the same moor the population level commonly fluctuates two or three-fold over a short period of years. The density is fixed at any time by the mean size of territory taken up by the cock birds in autumn and held from then until the breeding season is over by the following midsummer. In some winters the heather suffers from frost damage; some summers give better growth conditions than others. There were indications that changes in the heather were indeed affecting the birds, as suggested for example by sampling their body weights;

and that variations in the nutritive condition of the heather might explain why they defended bigger territories in some years than others. They might be taking bigger territories when food quality was poor, and smaller ones when it was good.

In an attempt to clinch the relationship between the consumer population and the state of their food resource an experiment was started in May 1965, on a very uniform tract of 32 ha of heather moor. It was divided into two halves, to one of which ammonium nitrate (with calcium carbonate) was applied as a fertilizer, in a single application giving 105 kg of nitrogen per hectare. Previous trials had shown that this would benefit the growth of *Calluna* and upgrade its nutrient quality. Before the experiment the breeding densities of grouse on the two parts were the same, namely four cocks and four hens. All these birds were shot during the summer so that none remained on the area. It was recolonized in the autumn territorial contest, five months after the fertilizer had been spread, by seven birds on each half instead of eight; but in the following summer of 1966 the breeding success was three times as good on the fertilized plot as on the control plot. In the autumn contest of 1966, roughly 17 months from the application of the fertilizer, the grouse count rose to 17 birds on the fertilized ground, but it remained unchanged at seven on the control half.

Measurement showed that the heather on the treated half had grown better and contained more nitrogen (Miller *et al.*, 1970), and if the two effects are multiplied together the rough approximation obtained suggests that, like the grouse population, the amount of nitrogen available in the young shoots per unit area had also increased between two and three times, compared with the control. The experiment has since been repeated on other areas with similar results, except that sometimes the number of grouse territories on the treated area has risen in the first autumn instead of the second one. There can be no doubt therefore that population density does reflect the nutrient state of the food supply, and that the earlier tentative conclusions of the research group (Miller *et al.*, 1966; Jenkins *et al.*, 1967) were correct.

After the chicks have hatched out, the territorial system, which has covered the moor in an almost continuous mosaic, breaks down. The adults then begin to moult and the young birds are growing fast. By September the young are the same size as their parents, and the family groups have broken up. The time approaches when a new territorial pattern will be set up, to last for another nine months. Young cocks may already challenge old ones, some of which have revived an interest in their previous territories. It is often in the earliest light of October mornings that the new disposition finally takes shape. With young cocks and old now on equal terms, the birds are spread out over the moor, challenging each other for the ground they occupy. Some hold their ground, and others give way; but to the end it is a sham fight, in which no-one gets seriously injured.

Always, at least in eastern Scotland, there are many unsuccessful males; in fact on average more than 60 per cent are unsuccessful. Although at first the new territories are defended only at dawn, and the outcasts can move about over the moor to feed during the rest of the day, some of them soon begin to lose condition, and they become an easy prey for predators. The established males accept the presence of some of the females, although they do not yet form lasting pair bonds, and they eject the rest, which join the outcast males and suffer the same fate. By February the established birds have paired, and they maintain their aggression against intruders for most of the day. The remaining outcasts are constantly harassed, and chased off the heather, so that many die of starvation. Their winter death rate from predation is seven times as high as it is in the established birds, and by the end of April virtually all of them are dead. With few exceptions, only established birds survive the winter.

Another important series of experiments showed that if territory-owning cocks were shot in autumn their places would be taken by the uppermost survivors among the outcasts. When all the birds present on a 12 ha area were shot in September they were replaced within a few days; whereas when the same was done in mid-December it was 2–4 weeks before the pattern of occupation was re-established. The density of re-colonization was very similar to the original density (Watson and Jenkins, 1968). This showed conclusively that the population density of established birds was being controlled by the territorial system, and that there were surplus birds available, even in December, that were capable of defending territories and subsequently breeding, provided they were given the chance.

The physiological basis of social success in the territorial contest was demonstrated by making experimental implants of an androgen pellet into single cocks that were caught for the purpose. Outcast males have smaller testes in winter than established males; but two outcast males implanted with pellets containing 15 mg of testosterone at the beginning of April forced their way even then into the territorial mosaic, obtained small territories, gained weight, and survived, although they did not obtain mates. A third bird, given an implant of corn oil as a control, was dead (as all three would otherwise probably have been) by the end of the month (Watson, 1970).

I have selected these items from the many results of the Grouse Research Unit's work as the most conclusive as far as my hypothesis is concerned. They leave no doubt that the grouse control their own population density, and that they vary the average size of the territories they take, from place to place and year to year, in conformity with the state of their food resource. The aggressive confrontation of the males is ritualized, and yet decisive. The result determines the quota that is to survive the winter and breed in the spring. At the same time it identifies the high proportion of losers as rejects, almost all of whom are doomed to be disposed of by predators, starvation, and disease. This verdict throws a novel light on Darwin's checks, as serving mainly in a secondary role by dispatching the victims of a prior social selection, conducted by the birds themselves.

Red grouse are exceptional birds only in minor respects. For example, for such large birds they are surprisingly fast-maturing and short-lived. Age brings no advantage to them in social status, so that when four months old they can compete on equal terms with their parents. They are so sedentary that immigration and emigration, as these are generally understood, enter very little into the homeostatic machinery. It has been suggested that their habitat in Scotland is to a considerable extent artificial, since it is maintained over much of the region by heather-burning, which prevents the regeneration of woodland and favours the fire-resistant *Calluna* on which the grouse feed. This is true enough, but has no bearing on the validity of the conclusions I have just drawn. Man may alter the habitat but he does not interrupt the homeostatic adjustment of the grouse. On the main research area no shooting is done except for occasional localized experiments; and on other moors where sport shooting does take place it seldom if ever removes enough birds, so that the homeostatic adaptations still discharge their function. The system has also been found to operate in very poor grouse habitats, for example in the West Highlands and in Ireland. The ptarmigan (*Lagopus mutus*), a closely related species, lives in a still-natural arctic–alpine environment at a higher elevation than that of the red grouse, and there its populations show a closely parallel system of homeostasis, also mediated by territoriality, and the exclusion of surplus birds (Watson, 1965). In short, although the red grouse habitat has been made more extensive and more uniform than it was in primeval times, the main effect, apart from increasing the number of birds, has probably been to intensify the link between them and their staple food *Calluna*, and to that extent to simplify the problems faced by the researchers.

I have not mentioned the variable reproductive output of the grouse. If one measures it in terms of the annual production of recruits per unit area, it depends first on the population density of breeding birds, and secondly on mean clutch size. The latter varies in correlation with the nutrient status of the heather. Reproduction normally budgets for a substantial surplus, which the habitat can easily carry in summer and autumn. The executive task of cutting recruitment down to the appropriate level is left until the autumn contest is over, and has identified the surplus birds.

The evolution of homeostasis

The symposium organizers asked me, in preparing this paper, to concentrate on the mechanics of population control rather than on the theoretical question of how the mechanisms evolved. What I have to offer on the evolutionary side is therefore just an outline.

In order to reap the benefits of population homeostasis, animals like the red grouse and beaver must be programmed to balance their population densities at levels that are safely within the carrying capacity of their habitats. This will ensure that, so far as their own actions can affect the future, their food-crops will continue to be renewed and will remain undiminished as succeeding generations pass. In each generation the consumers must not let their populations grow to the levels that could easily be attained, were it not for the conventional restraints they are programmed to observe; this in its turn implies that safeguards must exist to prevent individuals from breaking out from restraints in order to obtain selfish advantage in the short term.

There is a theoretical difficulty here which perhaps still needs to be clarified further. Fitness differs from one individual to another: we have seen for example that there are grouse which are successful in winning a territory and others which are not. Even among the successful birds the territories they defend differ considerably in size. There must be a block against selection moving the balance in favour of those birds that are willing to accept smaller territories; for if their preference was inherited and passed to their offspring, it would allow more of them to be accommodated with territories in any given area, and reduce the proportion that had to be expelled as outcasts. No harmful effect would be felt immediately. Generations might pass before the growth and biomass of heather had deteriorated enough to threaten the grouse stock with extermination.

Other self-regulated populations introduce a number of parallel theoretical problems. Sea-birds, for example, invariably nest in colonies (Wynne-Edwards, 1962). Feeding territories cannot be marked out on the sea-surface, but the birds hold nest territories instead, and they are strongly inhibited from breeding outside the conventional perimeter of an existing colony. In fact, so-called 'clubs' of non-breeding adults are to be found at the colonies of some species, and at least a proportion of the club members are potential nesters for whom there is no room inside the existing perimeter, although there is no lack of room of the same kind outside. New colonies are normally founded by break-away groups, rather than by solitary pairs.

Another example can be taken from the migratory locusts (Acrididae). Most species occupy more or less permanent bases called 'outbreak areas'. In years when there have been good rains, they reproduce very rapidly and build up huge travelling swarms of 'hoppers'. When these are adult, the migrant expeditionary forces become airborne and set off to invade, and presumably to colonize if possible, some more or less distant region. The invaders very seldom establish new permanent homes: instead, after at most a few generations, they die out leaving no descendants.

The resident stock that stays behind in the original outbreak area can well afford to produce these migrant swarms, because they have temporarily superabundant food which will be lost once arid conditions resume. The food consists of vegetation that will recover and produce crops on other similar occasions. The locusts' regime is adaptive, so long as new headquarter areas are found at least as often as old ones become uninhabitable. From the point of view of the genes, the strategy is prudential, though for decades or centuries at a time selection against the migratory habit is 100 per cent.

It was the consideration of such problems as these, while I was writing *Animal Dispersion*, that led me to think that there could be a process of group selection, which favoured local stocks whose members were adapted to accept restrictions placed on their own fitness, where history had shown these restrictions to be ecologically provident in the long term. Such stocks would be able to survive while others, less prudential, depleted their habitats and became extinct. The idea of group selection was by no means a new one, and it appeared to offer an alternative and complementary form of selection which could act on variations in hereditary traits, such as the size of territory to be claimed, or the reproductive rate, the cumulative effects of which could only be revealed and tested over periods much longer than the lifetimes of individuals. There were immediate critics at the time, pointing out that group selection was founded on fallacy. As a hypothesis it cannot be tested experimentally because of the prohibitive length of time that would be required; but in the last 15 years many theoreticians have wrestled with it, and in particular with the specific problem of the evolution of altruism. The general consensus of theoretical biologists at present is that credible models cannot be devised, by which the slow march of group selection could overtake the much faster spread of selfish genes that bring gains in individual fitness. I therefore accept their opinion.

Altruism in the evolutionary context means risking or sacrificing some of one's own fitness for the benefit of other members of the group. Thanks to the clear exposition by W. D. Hamilton (1964) it is now a familiar concept that, because an individual animal shares replicas of many of its genes with its relatives, it has in effect an inclusive fitness, vested not only in itself but also, to degrees depending on the closeness of the relationship, in all its kin alive at the time. It can be expected to act, therefore, in every way that will maximize this inclusive fitness; in a viscous type of population, where its companions are likely to include a number of relations, an individual may have opportunites of increasing the inclusive quantity by making measured personal sacrifices on their behalf. This is particularly easy to appreciate when parents engage in caring for their offspring. Genes for altruism could spread and evolve therefore in small sedentary populations, where the preconditions are easily met. The process has been called kin selection.

Altruistic adaptations are thus rather less paradoxical, in the eyes of the evolutionist, than prudential ones, using these two terms to differentiate between personal sacrifices that bring benefits to living contemporaries and those that bring benefits only to generations unborn. For each kind there is a secondary difficulty also, of how they can be maintained and protected from subversion by selfish mutations if, after the passage of time, the conditions which brought them into being no longer obtain: in particular, if the population regime should have changed and become less viscous. Nevertheless prudential adaptations, especially, appear to be common and widespread in the higher animals, regardless of major variation in population viscosity.

The answer to the riddle is probably shared between population structure and genetic mechanisms, in a manner still to be revealed. Deme structures are sufficiently common features of animal dispersion to suggest that they have an important adaptive function, which we are not at present able to take into account. Recessive mutations, when they first arise (and are not immediately lost again), make slow progress for many generations even under quite favourable selective pressures. Heterosis, the over-dominance of the heterozygote, is not an easy genetic mechanism for the neo-Darwinist to explain, although very widespread; it causes many of the fittest genotypes to segregate in gametogenesis, and that prevents them from being directly handed down to posterity. These are only some of the possible factors that might stand in the way of quick gains by selfish genes (cf. Lerner, 1953).

This symposium on population control by social behaviour is an indication of the interest the subject commands in the scientific world. Many biologists still think it is a very controversial subject, but fortunately there is no need to wait for someone to produce a model of how the system evolved before the facts themselves are assessed. Direct evidence that populations can be regulated through social or other intrinsic mechanisms has long been available, but in recent years it has greatly increased, and there are now convincing experimental results from mammals, birds, reptiles, fish, and a variety of insects. The symposium will add significantly to these results. What I hope it will also help to show is that doubt or controversy now clings only to the evolutionary mechanisms involved, something that would be equally true, for instance, of certain well-known genetic mechanisms. Population homeostasis is undoubtedly real, and its implications for animal ecology and human biology are much too great to be ignored.

References

Aleksiuk, M. (1968) Scent-mound communication, territoriality and population regulation in beaver (*Castor canadensis*). *Journal of Mammalogy*, **49**, 759–762.

Darwin, C. (1859) *The Origin of Species*. (References based on 6th Edition, with additions and corrections to 1872.) London: John Murray.

Green, H. U. (1936) The beaver of the Riding Mountain, Manitoba, an ecological study and commentary. *Canadian Field-Naturalist*, **40**, 1–8, 21–23, 36–50, 61–67, 85–92.

Hamilton, W. D. (1964) The genetical evolution of social behaviour. *Journal of Theoretical Biology*, **7**, 1–51.

Jenkins, D., Watson, A., and Miller, G. R. (1963) Population studies on red grouse (*Lagopus lagopus scoticus* Lath.), in north-east Scotland. *Journal of Animal Ecology*, **32**, 317–376.

Jenkins, D., Watson, A., and Miller, G. R. (1967) Population fluctuations in the red grouse (*Lagopus lagopus scoticus*). *Journal of Animal Ecology*, **36**, 97–122.

Lack, D. (1954). *The Natural Regulation of Animal Numbers*. Oxford: Clarendon Press.

Lerner, I. M. (1953) *Genetic Homeostasis*. Edinburgh and London: Oliver and Boyd.

Milkman, R. (1975) Specific death sites in a *Drosophila* population cage. *Biological Bulletin, Marine Biological Laboratory, Wood's Hole, Mass.* **148**, 274–285.

Miller, G. R., Jenkins, D., and Watson, A. (1966) Heather performance and red grouse populations. I. Visual estimates of heather performance. *Journal of Applied Ecology*, **3**, 313–326.

Miller, G. R., Watson, A., and Jenkins, D. (1970) Responses of red grouse populations to experimental improvement of their food. In *Animal Populations in Relation to their Food Resources*, ed. Watson, A. pp. 323–334. Oxford: Blackwell Scientific.

Moffat, C. B. (1903) The spring rivalry of birds: some views on the limit to multiplication. *Irish Naturalist*, **12**, 152–166.

Moss, R. (1967) Probable limiting nutrients in the main food of the red grouse (*Lagopus lagopus scoticus*). In *Secondary Productivity of Terrestrial Ecosystems*, ed. Petrusewicz, K. pp. 369–379. Warsaw and Krakow.

Watson, A. (1965) A population study of ptarmigan (*Lagopus mutus*) in Scotland. *Journal of Animal Ecology*, **34**, 135–172.

Watson, A. (1970) Territorial and reproductive behaviour of red grouse. *Journal of Reproductive Fertility*, Supplement 11, 3–14.

Watson, A. and Jenkins, D. (1968) Experiments on population control by territorial behaviour in red grouse. *Journal of Animal Ecology*, **37**, 595–614.

Wynne-Edwards, V. C. (1959) The control of population density through social behaviour: a hypothesis. *Ibis*, **101**, 436–441.

Wynne-Edwards, V. C. (1962) *Animal Dispersion in Relation to Social Behaviour*. Edinburgh and London: Oliver and Boyd.

Behaviour and population regulation in the Great Tit (*Parus major*)

JOHN KREBS and CHRISTOPHER PERRINS

Edward Grey Institute of Field Ornithology, Oxford, England

Introduction

The population ecology of the Great Tit (*Parus major*) has been studied in Oxford for the last 30 years, a project initiated by the late Dr D. Lack and carried out in the 600 acres of woodland in Wytham Estate. In this paper we will be reviewing some of the results of this long-term study, emphasizing in particular the role of behavioural interactions in determining population change at various times of year. We will also mention some relevant points about the population dynamics of the closely related Blue Tit (*Parus caerúleus*), and compare the results of the Oxford Great Tit study with a parallel study done in Holland largely under the direction of H. N. Kluijver. In comparing the two studies one should bear in mind that the birds in the two areas belong to different subspecies (*newtoni* in England, *major* in Holland) between which there are some striking contrasts in demography. Two facts are especially important: the Dutch birds are more migratory than the English, and they normally lay two broods in each season, while the English tits normally have only one brood.

The main advantages of the Great Tit as a study species are that it nests almost exclusively in boxes if these are provided in sufficient numbers, so that one can easily census the breeding population and production of young, and that the birds are easy to trap in winter at feeding stations. The provision of nest boxes probably alters the population ecology to some extent. In some studies it has been shown that the density of breeding pairs increases when boxes are provided (Enemar *et al.*, 1972) although this effect is much clearer in young plantations than in mature woodland with natural holes. Recent work has also shown that clutch size is smaller and breeding success lower in natural holes than in nest boxes (Nilsson, 1975; van Balen and Booy, 1976), and that clutch size varies according to the size of nest boxes (Löhrl, 1973). The provision of nest boxes therefore removes some variables from the dynamics of a natural population but, as far as we can tell, boxed and natural populations fluctuate in parallel, so the conclusions we draw from our study probably apply also to birds breeding in natural holes.

In table 1 we summarize the main features of the annual cycle of the Great Tit in Wytham, and indicate behavioural factors

Table 1 A summary of the major stages of the annual cycle of the Great Tit in Wytham Woods

Time of year	Stage of annual cycle	Possible adjustments of numbers
i Early spring (Jan–March)	Territories are set up	Some birds are unable to obtain territories, and therefore do not breed, at least in good habitats
ii Late spring (April–May)	Breeding season	Clutches vary considerably in size, fewer eggs being laid at high density
iii Early summer (June–July)	Young birds leave the nest	Within two weeks they are mostly independent of their parents. Aggression between young birds is common; many young birds disappear
iv Early autumn (Sep.–Oct.)	Resurgence of territorial behaviour	Numbers remaining in an area may be adjusted. However, many young birds are not apparently involved, at least in English populations
v Late autumn (Nov.)	Winter population more or less stabilized	A very variable proportion of the summer's young are still in the population at this stage
vi Winter (Dec.–Jan.)	Shortest days for feeding. Birds moving around in mixed flocks. Some territorial behaviour in mild weather	Some mortality, but usually less than that preceding stage v

which might cause changes in population size. Starting with the breeding season, we can identify three periods in the year during which behaviour may play a role in limiting population density. Clutch size decreases as density increases, possibly as a result of behavioural interactions; the disappearance of many young birds in the summer and early autumn coincides with an increase in aggressive encounters, either in the form of dominance hierarchies among fledglings or autumn territorial activity largely by adults; finally, the spring breeding density is probably determined by territorial behaviour.

Before we go on to discuss each of these periods in more detail, it is important to clarify two points of terminology. When we refer to the 'population' of Great Tits, we mean the birds living in our study area rather than an isolated unit. As it turns out, immigration and emigration are among the important processes influencing the population in our area, so strictly we should talk about the limitation of local population *density* rather than of total population *size*. We should also mention that our study population is living in an optimal habitat (mature mixed oak woodland), and it is possible that the factors influencing reproduction and mortality are different in suboptimal habitats such as gardens (Perrins and Moss, 1975; Lack, 1966). A second point is our use of the term 'regulation'. Here we follow the usual tradition of using the word to refer only to density-dependent, stabilizing influences on the population. In general, the action of density-dependent processes can be demonstrated either by analysis of population censuses, for example by k-value analysis (Varley *et al.*, 1973), a method which presents both logical (Murdoch, 1970) and technical (e.g. Ito, 1972) problems, or by perturbation experiments in which the population is artificially increased or reduced in size, and observed to return to the same level.

Reproductive output and density

Both in England (Lack, 1966) and in Holland (Kluijver, 1951) the production of young per pair is density dependent. This results from two main effects: clutch size declines with increasing density, and the proportion of pairs attempting to rear second broods declines with increasing density. This latter effect is more marked on the continent than in Britain, where rather few birds have two broods even in low density years. In Wytham there is also a third density-dependent component of reproductive output, namely predation by weasels on eggs and early young (Krebs, 1970). Dunn (1977) has recently confirmed and extended this finding in a more detailed analysis. He also shows that the extent to which weasels prey on tit nests depends not only on density, but also on the availability of an alternative prey in the form of rodents. The two factors contribute about equally in explaining variations in the rate of attack on tit nests.

Density dependence of clutch size

We know very little about the mechanism producing a density-dependent reduction in clutch size, but the available evidence seems to suggest a behaviourally mediated response, perhaps analogous to the effects on fecundity, documented in artificially crowded rodent populations (Christian and Davies, 1964). Two simple hypotheses about the proximate mechanism of reduction in clutch size are either that the female is more likely to run short of food during egg laying when the population density is high, or that even though food is plentiful the female responds to behavioural cues such as increased frequency of singing in the area or rate of encounters with other birds.

The first of these alternatives seems unlikely, since several pieces of evidence suggest that females are not limited by the quantity or quality of food at the time of egg-laying: a bird which loses its entire clutch to a predator will lay another one of a similar size almost immediately; at the time when the female stops laying the food supply is rapidly increasing (Perrins, 1965); and experimentally increasing the food supply at the time of egg-laying results in clutches being started earlier in the season, but not in them being larger (Källander, 1974). In an experimental study of aviary populations of starlings (*Sturnus vulgaris*), Risser (1975) showed a density-dependent reduction in clutch size in birds supplied freely with food.

Hence, it seems likely that the reduction in clutch size is brought about by a behavioural–physiological mechanism rather than directly by nutritional state of the female. The ultimate advantage to a female of laying a smaller clutch at high densities is that it enables her to produce heavier young which will have a chance of surviving in spite of increased competition. Survival of young soon after fledging is related to weight, heavier chicks being more likely to survive (see below), and in years of high breeding density when many young are produced, the summer competition between young to survive is more intense, strongly favouring heavier young. Other things being equal, chicks from smaller broods are heavier and hence survive better.

Proportion of second broods

On the Continent, as already mentioned, the females are more likely to start second broods in years of low than of high density. The young from second broods survive less well in competition with the young of first broods. If the latter are removed, the survival of the former is increased (Kluijver, 1971) (see below). In the natural situation, the young of second broods tend to leave the breeding area and wander further afield than the young of first broods, apparently because they cannot easily establish themselves in competition with first broods (Kluijver, 1971). One can again suggest an evolutionary explanation for the variation in the proportion of second broods, namely that the higher the population density, the lower the chance that second broods will produce surviving young and hence the less likely the females are to have them.

The evidence would therefore suggest that the factors responsible for variation in reproductive output are behavioural and, although they can be explained in terms of evolutionary advantage, the exact mechanisms remain unknown.

Survival of young

In Wytham, each pair of Great Tits rears an average of about six young. Ringing studies have shown that only about one (17 per cent) of these young survives to the following breeding season; this, together with a 50 per cent adult survival rate, means that the population is roughly stable (Bulmer and Perrins, 1973). This is, of course, only an average picture, and the breeding population fluctuates from year to year by as much as twofold. These annual changes in breeding density result mainly from changes in the survival of young birds (Perrins, 1965), the survival of adults being remarkably constant from year to year. Thus it is crucial to the understanding of population change to know what happens to the young birds which disappear fairly soon after fledging.

Two main arguments have been put forward to account for this disappearance. One suggests that the crucial stage for survival of young is during the autumn resurgence of territorial activity. Kluijver (1951, 1971) in Holland and Dhondt (1971) in Belgium have suggested that young birds are forced to emigrate from their area of birth during the period of autumn territorial behaviour and that second-brood young are less likely than first broods to succeed in establishing a territory. In England, the same argument could be applied to dominant and subordinate young from the single brood normally reared. The second suggestion is that young birds often die of starvation just after leaving the nest, the survival at this stage being very variable, but once birds have survived this critical period they have a high chance of becoming members of the breeding population (Perrins, 1965). These theories are not mutually exclusive; there may be a reduction in the number of young both immediately after fledging in June–July, and later in September–October.

The timing of disappearance of young

The most direct evidence concerning the timing of juvenile disappearance comes from analysing the juvenile : adult ratios among birds captured at different stages during the summer and autumn. Figure 1 shows the results of this analysis for the Great Tit population in Wytham (Webber, 1975). The 'autumn territory' hypothesis would predict a sudden drop in the juvenile : adult ratio in September as young birds are forced to emigrate, while the 'starvation' hypothesis predicts a sudden drop in June–July, after fledging. The data do not support one alternative to the exclusion of the other: instead there is a steady drop in the proportion of juveniles through the summer and autumn, perhaps suggesting that both effects operate. After October, there is little or no differential mortality of adults and

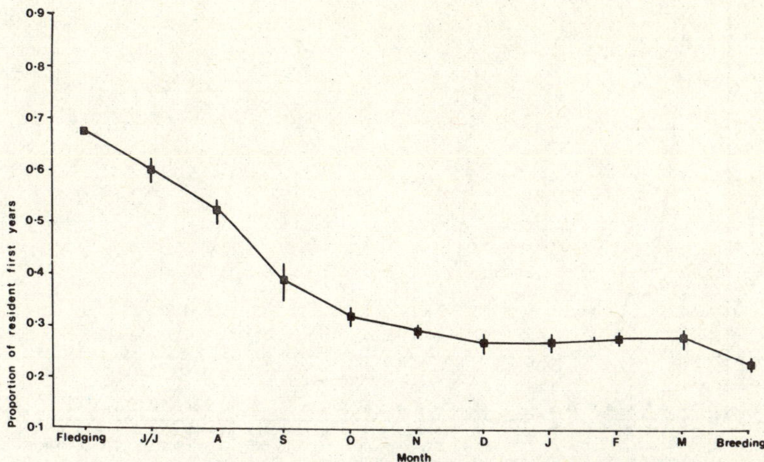

Figure 1 Changes in the age ratio of resident Great Tits trapped in Wytham Woods at different times of year. Data are averaged over the period 1961–1974 (from Webber, 1975). Immigrant birds are excluded from this graph, but there is an autumn peak of immigration in October.

juveniles, so that the critical mortality or emigration producing fluctuations in numbers (not necessarily regulation) occurs before November.

A similar analysis of much more extensive data on the Blue Tit (*Parus caeruleus*) is shown in figure 2 (Perrins, in preparation). This species is a close relative of the Great Tit and usually shows parallel fluctuations in numbers, so we are almost certain that the factors influencing juvenile survival of the two species are similar. As with the Great Tit, the proportion of young birds in the population declines steadily from the time of fledging to November, and then remains more or less constant (figure 2). The data for Blue Tits, which were obtained from ringing schedules of the British Trust for Ornithology, have a bias towards juveniles and so must be interpreted with caution. The ratio of juveniles to adults at the time of fledging should not be higher than about 4:1 as each pair rears eight young on average. When the ringing records were analysed with respect to habitat, there was no clear evidence for movement into marginal habitats such as gardens, which one might expect if young are evicted from optimal areas; the age ratios tended to change more or less synchronously in all habitats, although

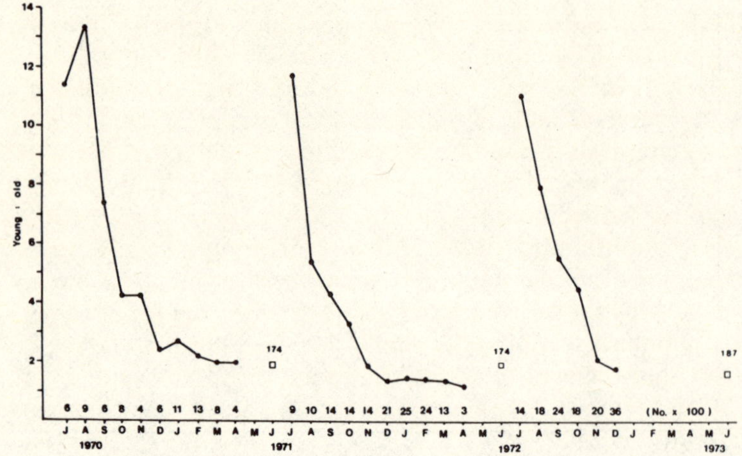

Figure 2 Changes in the age ratio of Blue Tits with season in three successive years, from analysis of British Trust for Ornithology ringing records (excluding retraps). The open square shows the breeding season age ratio for Wytham Woods, as there are insufficient data from B.T.O. records for this time of year (from Perrins, in prepn).

some habitats such as reed beds held high numbers of juveniles in late summer.

Starvation or emigration?

Perrins (1965) showed that the young birds surviving through the summer tend to be heavier than those which disappear, and at least some of the disappearance of light birds happens within two or three weeks of leaving the nest (although Webber (1975) was able to confirm this for Great Tits in only one of two seasons). The heavier birds are, however, not merely those with more stored food in the form of fat, but they are ones with larger body size excluding fat (Garnett, 1976). This suggests that the mechanism underlying the disappearance of light birds is not, as Perrins originally suggested, shortage of food. Garnett (1976) found that even within a few days of fledging, dominance, hierarchies developed among captive broods of Great Tits and heavier birds tended to be dominant, which suggests a behavioural mechanism for the differential survival of young.

The factors influencing the disappearance of young birds in late summer have been studied experimentally by Kluijver (1966, 1971) in Holland. On the island of Vlieland, he showed that the survival both of young and especially of adults was greatly increased if he culled 60 per cent of the young. He also found in a mainland population ('Hoge Veluwe') that second-brood birds settled nearer their birth place after the removal of 90 per cent of the first broods. Kluijver interprets these results by suggesting that normally the second-brood young are forced to emigrate by the autumn territorial activity of the first broods, which establish territories earlier. The increase in survival of adults after removal of young is more difficult to explain as adults are known to dominate young birds in struggles for territories. As Kluijver (1971) shows, the adults did not survive better simply because they expended less energy in rearing young. It is also worth pointing out that although Kluijver's experiments suggest that emigration of young is density dependent (in fact he has shown that emigration is dependent on the density of both juveniles and adults), Krebs (1970) and Webber (1975) were unable to find any evidence of density-dependent emigration of young birds from Wytham.

One further point must be made about the 'disappearing' young. They cannot be emigrating and surviving well elsewhere. At least the major changes in population size occur synchronously over extensive areas of western Europe. Hence surplus young produced in a good year in one area cannot move to another area since that, too, will have a high population. The large numbers of young that are 'excluded' from an area must mostly die. This is not to say that there are no local adjustments of numbers. Birds may tend to congregate in areas with a rich food supply and settle nearby in the spring.

Relationship to the Beech crop

Neither of the two main hypotheses outlined above have as yet been able to explain at all adequately what environmental factors lead to a highly variable number of the young being able to settle in their natal area in different years. Only one striking correlation between the environment and the numbers settling is known, the size of the beech crop. If the latter fails, the proportion of young surviving to breed will be low; if there is a large crop, the numbers settling will be large (Perrins, 1966). There are, however, at least two reasons why this correlation cannot be wholly causal. Firstly, the tit populations fluctuate in parallel with the beech crop in areas where there is no beech. There is, in fact, likely to be a simple explanation for this: many other tree species tend to crop heavily in years when beech mast is abundant; hence beech can probably be regarded as an indicator of the general seed-crop levels. More puzzling, the tits do not seem to feed heavily on the seeds until late autumn, by which time much of the differential survival of the young appears to have occurred. The explanation of this observation may lie along the same lines as that given above, namely that the tits find other foods earlier in the autumn that fluctuate in parallel with the beech mast. At present this matter remains unresolved, as does the mechanism whereby the tits' numbers are adjusted in relation to the seed crop.

In summary, we know that the survival of young birds between fledging and autumn is important in causing population change. Heavier (but not fatter) birds are more likely to survive; and aggressive interactions between young from July onwards may play a role in differential survival. This proximate behavioural mechanism is probably linked to food supply as indicated by a correlation between survival and the size of the beech crop. Autumn territorial behaviour is more important in the exclusion of young in Dutch populations than in English. This difference may reflect the fact that Dutch birds rear getting on for twice as many young as those in Wytham, so that competition between young is more severe in Holland.

Winter disappearance and spring breeding density

Territory and breeding density

As we described earlier (table 1), Great Tits in Wytham usually live in mixed-species flocks between November and mid-January, when they begin to establish spring breeding territories. This general pattern is flexible; in mild seasons adult birds may be territorial more or less throughout the winter, while a cold spell in March may cause birds to leave their territories and join

Table 2 Summary of removal experiments carried out in Wytham

Date		Birds removed		Replacements	
		male	female	male	female
March 1968	1	7	7	8	5
March 1969	1	7	7	7	7
Feb. 1975	2	8	8	8	8
March 1975	2	8	8	4	4
Feb. 1976	3	8	8	4	4
March 1976	3	4	4	2	2
TOTAL (pre-breeding)		42	42	33	30
May 1973	4	4	9	0	0
May 1974	4	6	7	0	0
TOTAL (breeding season)		10	16	0	0

Sources: 1 Krebs (1971); 2 Krebs (1977a); 3 Krebs (unpublished); 4 Webber (1975).

again into flocks. Kluijver and Tinbergen (1953) first made the suggestion that spring territoriality might have the effect of limiting the breeding density of Great Tits in optimal habitats. Their evidence was rather indirect: they noted that the breeding population was more constant in good habitats (oak woodland) than in neighbouring poor areas (pine woodland), and suggested that birds compete first for territories in the former areas, and later 'surplus' birds, having failed to get a territory in the best habitat, spill over into other areas. The size of the 'surplus', hence the size of the breeding population in poor areas, varies from year to year.

The hypothesis that territoriality limits the breeding density was tested experimentally by Krebs (1971), who plotted the territories in a 15 ha piece of mixed oak woodland during early spring, and, once the boundaries were established, removed half the resident pairs. The experiment was done in each of two seasons, and involved in total the removal of 28 birds (14 pairs). The empty spaces were rapidly (within a few days) refilled with new pairs (table 2) as predicted by Kluijver and Tinbergen's hypothesis. There is considerable competition for good territories, and some pairs are excluded from the optimal breeding habitat. Some of the replacement pairs were shown to have come from farms, gardens, and orchards surrounding the oak woodland, and these habitats are known to be relatively poor in terms of breeding success (Lack, 1958; Perrins, 1965; Krebs, 1971) (although it is possible that the birds have a greater longevity in agricultural and suburban habitats where winter food is probably more plentiful). The conclusion that territoriality limits the breeding density applies not only to Great Tits: removal experiments on a variety of other birds (e.g. Watson and Moss, 1972; Knapton and Krebs, 1974; Harris, 1970) have demonstrated the same effect.

Subsequent removal experiments with Great Tits have in part confirmed the earlier conclusions (Krebs, 1977a, unpublished). In 1975 and 1976 a total of 16 pairs and 12 pairs respectively were removed from a 6 ha mixed oak wood, and were rapidly replaced with newcomers (table 2). The density of birds was lower after the removals than before, but the territories of newcomers completely filled the wood. These more recent experiments demonstrated several additional points. Firstly, in both years, two successive removals were done in the same piece of woodland and, although the breeding density was considerably lower at the end of removals than at the beginning, the fact that there were still some 'surplus' pairs available to reoccupy territories after the second removal, is a further indication of the importance of territorial exclusion. The second point is that the replacement birds did not, as Krebs (1971) had reported, come mainly from farmland and garden territories. In fact the origin of most of the birds was unknown: some had been caught in the wood at feeding stations earlier in the winter; others were unringed birds, apparently not regular visitors to the wood.

Table 2 also shows the results of removal experiments carried out later in the spring, after the start of breeding (Webber, 1975). 16 females and 10 males were captured during the incubation stage of the season. None of these were replaced, which suggests that 'surplus' birds either give up or establish territories elsewhere. Eyckerman (1974) and Drent (1976) have reported that surplus birds on the Continent sometimes manage to breed inside the territories of residents. We have never observed this in Wytham: perhaps the small size of the territories in our study area compared with those in Holland

and Belgium enables residents to detect intruders more readily and evict them.

An alternative experimental approach in studying the role of territory in limiting breeding density is to add birds to the spring population, the prediction being that few, if any, will succeed in settling, and that the breeding population will stay close to normal. This experiment was done by Webber (1975) who added a total of 36 birds to the population of Bean Wood in spring 1973, 57 in late 1973–early 1974, and 13 in 1975 (table 3). Only eight out of the total of 106 birds settled, though there was a slightly higher chance that birds would settle if released in the winter and early spring than if released later when territory boundaries are well established. Although there are some difficulties in the interpretation of this experiment (for example, released birds may have tried to home to their capture point about five miles away, although they were not found there), it supports the idea that spring territories limit the breeding density in optimal habitats.

Webber (1975) also recorded the pattern of settling of birds in woodland and hedgerow territories to see if the optimal areas are filled up before birds settle in suboptimal habitats, as reported by Kluijver and Tinbergen (1953). Figure 3 shows the pattern of settling in and around Bean Wood in 1974. The upper graph shows the proportion of all (young and adult males)

Table 3 Results of an experiment in which male and female Great Tits were released in Bean Wood during 1973, 1974, and 1975 (Webber 1975)

Release date	Number released		Number settling on territories in wood	
	male	female	male	female
Feb. 1973	10	15	3	0
March 1973	4	7	0	1
Nov.–Dec. 1973	12	8	0	1
Jan. 1974	7	7	0	1
Feb. 1974	4	9	0	0
March 1974	5	5	0	0
Feb. 1975	3	5	1	1
March 1975	3	2	0	0
Winter–early spring	80		7	
Late spring	26		1	

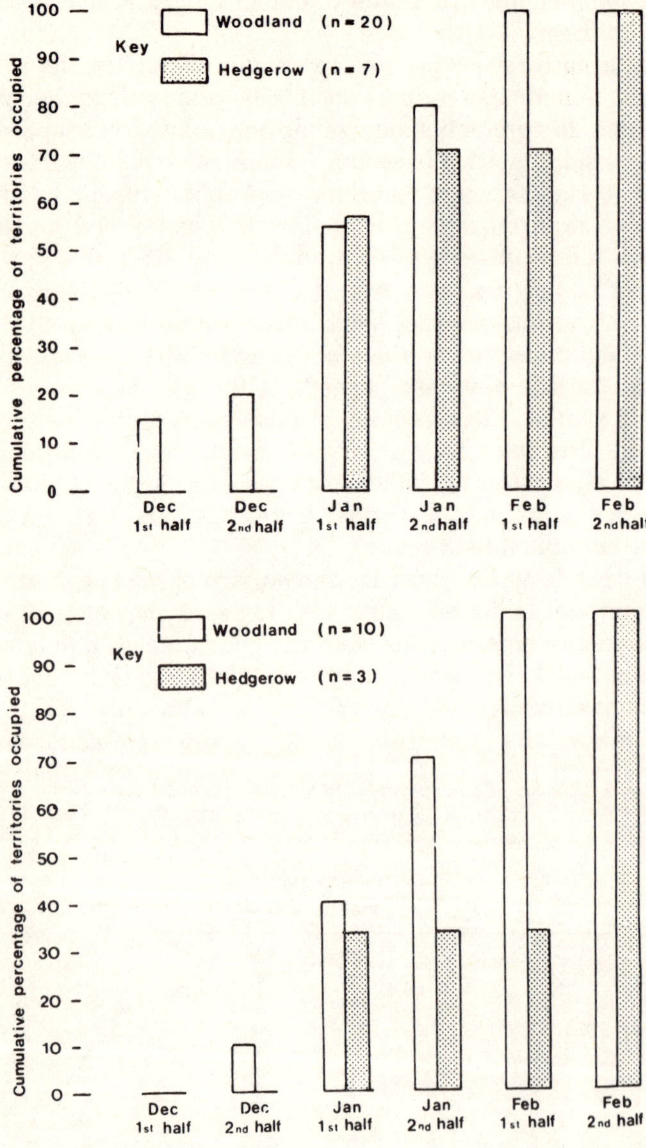

Figure 3 The settling pattern of birds in and around Bean Wood in 1974. The upper graph shows adult and first-year males combined, the lower graph first-year birds only (from Webber, 1975).

territories occupied at different stages during the winter and early spring, while the lower graph refers to first-year birds only. Young birds, but not adults, appear to settle earlier in woodland, and only occupy the marginal territories when the woodland is full, although the sample size is rather small. However, young birds do not settle appreciably later than adults, which suggests that the competition for territories in optimal habitats starts before the birds settle on territories, perhaps during the period of winter flocking.

In support of this, Webber and Krebs (unpublished) found that birds subsequently establishing territories in a mixed woodland tended to be dominant in winter food fighting at feeding stations (see also Knapton and Krebs, 1976). Since the answer to the question of why some birds and not others succeed in establishing territories in woodland is in part related to dominance hierarchies in winter flocks, we are led to ask what makes a bird dominant. One important factor is age: Krebs (1971, 1977a) found that replacement males are younger than removed territory holders. Garnett (1976) showed that heavier males tend to be dominant at winter feeders, suggesting that, at least during winter, heavier males are at an advantage (see also section on late summer arrival).

Although weight and age and therefore dominance in winter flocks is one factor determining whether or not a male succeeds in securing a territory in optimal habitats, autumn competition for territories also plays a role. Some of the birds defending spring territories, including all the surviving adults from the previous season, have already established territories in the early autumn (as in Holland—Kluijver, 1951) which they may leave during the winter and reoccupy in early spring. In fact it is difficult to decide on present evidence whether high dominance status in winter results from, rather than leads to, establishment of territories. A further complication is that dominance relationships between territorial and 'surplus' birds can be changed experimentally. In two experiments, Krebs (1977a, unpublished) captured a total of twelve territorial pairs in March 1975 and 1976 (see table 2) and, after new birds had completely occupied the wood, he released most of the original territory owners (10 pairs). Only one pair succeeded in displacing a newcomer and re-establishing a territory, and at least two pairs bred in suboptimal habitats. Thus prior occupancy can play a major role in determining which birds succeed in territorial competition.

Behavioural mechanism of territory maintenance

The removal and addition experiments show that territoriality limits breeding density, but do not show exactly how 'surplus' birds are excluded or, to put it another way, which aspects of the residents' behaviour are used by intruders in selecting whether or not to settle. This question has been studied by Krebs (1976, 1977a) who removed territorial pairs and placed loudspeakers in their empty territories, broadcasting the songs of the former residents. Invaders are, at least in the short term, discouraged from trying to settle in the area. Thus one cue used by invaders in choosing where to settle is the song of resident birds. Krebs (1977b) also suggests that resident birds may use their varied song repertoires to further discourage invaders from settling, by creating an illusion of a densely crowded habitat with many resident territory holders—the so-called Beau Geste effect.

Great Tits, in common with many other songbirds, show most intensive territorial activity just after dawn, and this is especially true of song (table 8, Hinde, 1952). One explanation of this is that invaders are more active in searching out and competing for territories in the early morning than at other times, an interpretation favoured by the results shown in figure 4: new birds are much more likely to settle on empty territories after the removal of residents during the early morning than at other times of day. The vertical axis of figure 4 shows the rate of arrival of new pairs into a 6 ha woodland, after all the residents were removed, measured by dawn to dusk observation until all territories were occupied.

Annual changes in average territory size

Within any one year, territorial behaviour seems to limit the spring breeding population of optimal habitats in a density-dependent manner (figure 5), and yet territory size is so variable from year to year that it has little density-dependent or regulatory influence on the Great Tit population over a longer

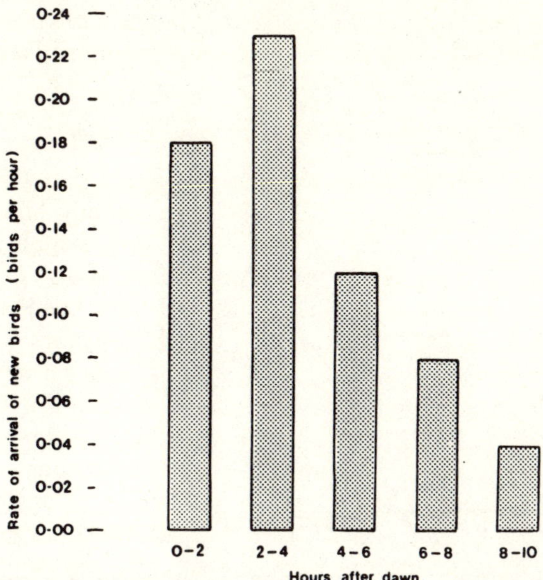

Figure 4 The rate of arrival of new birds into empty territories at different times of day (from Krebs, in prepn).

time scale (Lack, 1966; Krebs, 1970, 1971; Webber, 1975). In other words territory only has a density-dependent influence within limits set by year-to-year changes in some other feature of the population or environment. As we have discussed in the previous section, the main factor causing annual fluctuations in population size is survival of young birds, so the question arises whether spring territory size is largely set by the number of survivors from the previous year, or whether it is adjusted to winter resources.

The most obvious environmental resource to which territory size could be adjusted is winter food supply. Although Krebs (1971) found no increase in breeding density (decrease in territory size) after supplying a food supplement to the tit population during one winter in a 15 ha wood, both von Haartman (1973) in Finland, and van Balen (1976) in Holland have shown a marked increase in breeding density after winter feeding. The latter two studies were done over several years and so the results may be more reliable than the one year study of Krebs. However one should also bear in mind that the natural

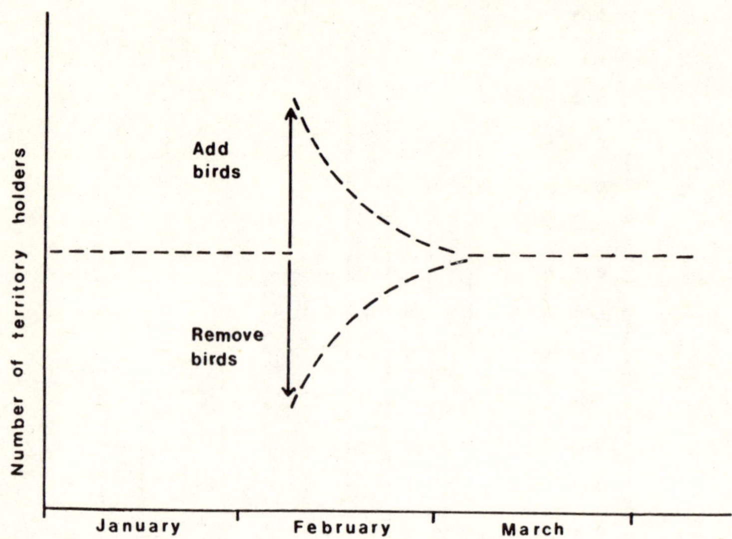

Figure 5 A summary of the experimental manipulations showing density dependence, within one season, of spring territoriality.

winter food supply is almost certainly much more limited in Finland and Holland than in England. How does winter food supply influence territory size? One possibility is that individuals directly adjust their territories in relation to food availability (e.g. Watson and Moss, 1972; Stenger, 1958; Simon, 1975; Slaney and Northcote, 1974). More likely in the case of the Great Tit is that additional winter food increases the survival of adults (van Balen, 1976) and/or juveniles, as well as encouraging some immigration, and this in turn leads to greater pressure on territories, forcing residents to contract their defended areas. On this argument, the proximate factor causing year-to-year changes in territory size is simply pressure from surplus birds, and the ultimate factor is food, or whatever other environmental change causes a change in survival and immigration. Pressure from neighbours does influence territory size in the Great Tit: birds expand their territories when neighbours are removed, and birds at the edge of a wood tend to have larger territories and fewer neighbours (Krebs, 1971).

There are two other proximate factors which could influence year-to-year changes in territory size, but probably neither are as important as intruder pressure. One is age: Dhondt (1971) reported that first-year males defend smaller territories than adults and suggested that a change in the age structure of the population may alter the spring breeding density, a rather circular argument since high breeding populations have to contain a lot of juveniles. However, the effect of age on territory size in Wytham seems to be small or non-existent (Krebs, 1971, 1977a; Webber, 1975). Webber found that young males have smaller territories than older birds only in a year when they settled later, and as the young birds in Dhondt's study area also settled later than adults, it is possible that the sequence of settling influences territory size. Van den Assem (1967), Krebs (1971), and Maynard Smith (1974) suggested that the temporal pattern of settling on territories could influence territory size, and the idea has been confirmed experimentally by Knapton and Krebs (1974). These authors removed territorial song sparrows (*Melospiza melodia*) in two areas; in one area they took away the residents one pair at a time, while in the other area they removed all the residents at once. The conclusion was that in the latter area synchronous settling by new birds gave rise to decrease in territory size, while asynchronous settling in the other area resulted in no change in average territory size. While this effect may play a role in tit populations, it is unlikely to have a major influence since it only produced a 50 per cent increase in population density in the study of Knapton and Krebs, while the breeding density of Great Tits can change by as much as 100 per cent from year to year.

To sum up this section, breeding density in optimal habitats is limited by territorial behaviour. Dominant birds in winter flocks are heavier than average and tend to become territory holders in the spring although this may be an effect rather than a cause. Year-to-year changes in territory size, while ultimately linked to some resource such as food, may be proximately caused by pressure on territories from young birds.

Discussion

In summary, the main conclusions we can draw about the role of behaviour in influencing population density in the Great Tit are as follows: density-dependent changes in clutch size are

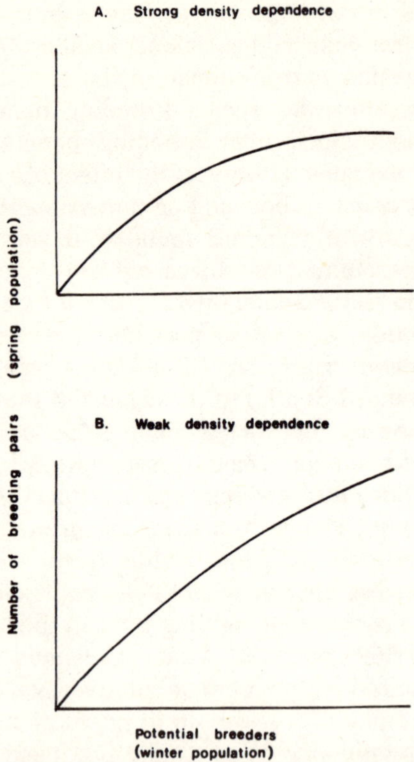

Figure 6 The influence of territorial behaviour between seasons is not strongly density dependent (A), but acts more in the manner shown in B (from Krebs, 1971).

possibly brought about through a behavioural mechanism; body-size dependent dominance relationships contribute to the disappearance of young birds in the summer and early autumn; and spring territoriality has a density-dependent effect in determining breeding density *within* a year. This effect is weak *between years* probably because the extent to which territories are compressed depends on the number of young birds surviving through the autumn (figure 6).

Although we assume that these behavioural factors are proximate mechanisms of population limitation ultimately geared to some environmental resource, we have virtually no concrete information about the critical resource(s) involved. Survival of juveniles is well correlated with the size of the beech crop (Perrins, 1966; Webber, 1975) but, as we explained earlier, this cannot always be a direct causal link. Food availability is implicated in winter survival by the feeding experiments in Finland and Holland, but not in England, so at the moment this important aspect of population regulation remains unresolved.

As we mentioned in the introduction, one method of investigating the role of mortality at different times of year in the regulation of population density is by k-value analysis or similar regression techniques. This sort of analysis has been done on the Great Tit data by several authors, notably Pennycuick (1969), Krebs (1970), and Webber (1975). Webber's analysis, and model based on the analysis, is the most recent and detailed. He divided the components of population change outside the breeding season into juvenile mortality and migration, first-year immigration and adult mortality. These components were analysed as k-values in a stepwise multiple regression analysis to examine for relationships between density, as well as numerous environmental factors and the beech crop index. Webber's main conclusion was that the only important density-dependent effect outside the breeding season was late summer and autumn immigration of first-year birds, especially females (figure 7). The immigration rate was also related to the size of the beech crop. This result is rather surprising, but it fits in with our earlier suggestion that autumn dominance or territorial interactions may play a major role in regulating population density. The surprising thing is that immigration, rather than emigration, is density dependent, as it implies that the source population of the immigrants is regulated in a different way from the Wytham population itself.

Krebs (1970) proposed that two of the density-dependent components of reproduction, reduction in clutch size and predation by weasels, could be sufficient to regulate the population. Table 4 shows a simple example of how a small change in clutch size with density could have a depressing effect on the population which is proportional to density. However, Webber (1975) discusses these weak density-dependent effects, and demonstrates in a simulation model that the return time of the natural population after a perturbation in density is too rapid to be produced by density dependence of

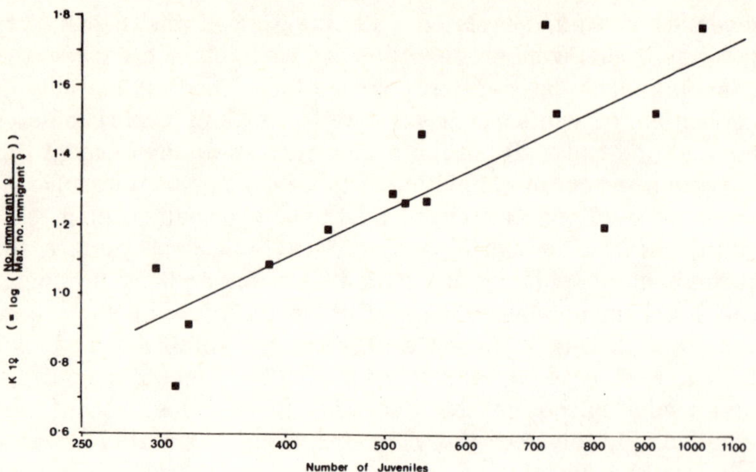

Figure 7 Immigration of the first-year birds, especially females, is the main density-dependent effect operating outside the breeding season, according to the analysis of Webber (1975).

reproductive output. In order to generate the observed pattern of population change, Webber had to incorporate into his model density-dependent immigration of young females in the autumn. Thus behavioural interactions in the late summer or early autumn probably play a crucial role in regulating population density.

We have suggested that territoriality in the spring and aggressive encounters between young birds in the late summer have major effects on population density. However, territoriality evolved as a mechanism of individual competition for resources, and its effects on population regulation are incidental. We have not yet identified all the selection pressures acting to favour

Table 4 An example to illustrate how density-dependent changes in clutch size could bring about regulation of the population, assuming adult mortality (50%) and juvenile mortality (90%) are both constant

Population size (pairs)	Summer total (adults + eggs)	Total next year (50% adult mortality, 90% juv. mortality)	Per cent change
80	160 + 640	80 + 64 = 144	−10
10	20 + 110	10 + 11 = 21	+5

territoriality in the Great Tit, although Krebs (1971) showed that more widely spaced nests have a lower chance of succumbing to predation by weasels.

Acknowledgements
J. K. was supported by the Science Research Council. We thank Maggie Norris for typing the manuscript and Ruth Ashcroft for drawing the figures.

References

van den Assem, J. (1967) Territory in the Three-spined Stickleback *Gasterosteus aculeatus*. *Behaviour*, Supplement, 16, 1–164.

van Balen, J. H. (1976) Population dynamics of the Great Tit *Parus major*. 1. Factors affecting the size of the breeding population. *Verhandelingen Koninklijke Nederlandse Akademie Wetenschappen Afdeling Natuurkunde 2ᵉ Reeks*, **67**, 3–5.

van Balen, J. H. and Booy, K. (1976) Population dynamics of the Great Tit *Parus major*. 5. The breeding of Great Tits in natural nest sites. *Verhandelingen Koninklijke Nederlandse Akademie Wetenschappen Afdeling Natuurkunde 2ᵉ Reeks*, **67**, 11–12.

Bulmer, M. G. and Perrins, C. M. (1973) Mortality in the Great Tit, *Parus major*. *Ibis*, **115**, 277–281.

Christian, J. J. and Davies, D. E. (1964) Endocrines, behaviour and population. *Science*, **146**, 1550–1560.

Dhondt, A. A. (1971) The regulation of numbers in Belgian populations of Great Tits. In *Proceedings of the Advanced Study Institute on 'Dynamics of Numbers in Populations'*. pp. 532–547. Oosterbeek, 1970.

Drent, P. (1976) Population dynamics of the Great Tit *Parus major*. 2. Territorial behaviour and population dynamics. *Verhandelingen Koninklijke Nederlandse Akademie Wetenschappen Afdeling Natuurkunde 2ᵉ Reeks*, **67**, 6–8.

Dunn, E. K. (1977) Predation by weasels (*Mustela nivalis* L.) on breeding tits (*Parus* spp.) in relation to the density of tits and rodents. *Journal of Animal Ecology*, **46**, 633–652.

Enemar, A., Nyholm, E., and Persson, B. (1972) The influence of nest-boxes on the passerine bird community of Fagelsangsdalen, Southern Sweden. *Vår Fågalvärid*, **31**, 263–268.

Eyckerman, R. (1974) Some observations on the behaviour of intruding Great Tits *Parus major* and on the success of their breeding attempts in a high density breeding season. *Le Gerfault*, **64**, 29–40.

Garnett, M. C. (1976) Some aspects of body size in tits. D.Phil. thesis, University of Oxford.

von Haartman, L. (1973) The population of the Great Tit at Lemsjöholm. *Lintumies*, **1**, 7–9.

Harris, M. P. (1970) Territory limiting the size of the breeding population of the Oystercatcher (*Haemotopus ostralegus*)—a removal experiment. *Journal of Animal Ecology*, **39**, 707–713.

Hinde, R. A. (1952) The behaviour of the Great Tit (*Parus major*) and some related species. *Behaviour*, Supplement 11, 1–201.

Ito, Y. (1972) On the methods for determining density dependence by means of regression. *Oecologia (Berl.)*, **10**, 347–372.

Källander, H. (1974) Advancement of laying of Great Tits by the provision of food. *Ibis*, **116**, 365–367.

Kluijver, H. N. (1951) The population ecology of the Great Tit, *Parus m. major* L. *Ardea*, **39**, 1–135.

Kluijver, H. N. (1966) Regulation of a bird population. *Ostrich*, Supplement, 6, 389–396.

Kluijver, H. N. (1971) Regulation of numbers in populations of Great Tits (*Parus m. major*). *Proceedings of the Advanced Study Institute on 'Dynamics of Numbers in Populations'*. pp. 507–523. Oosterbeek, 1970.

Kluijver, H. N. and Tinbergen, L. (1953) Territory and the regulation of density in titmice. *Archives néelandaises de zoologie*, **10**, 265–289.

Knapton, R. W. and Krebs, J. R. (1974) Settlement patterns, territory size, and breeding density in the song sparrow (*Melospiza melodia*). *Canadian Journal of Zoology*, **52**, 1413–1420.

Knapton, R. W. and Krebs, J. R. (1976) Dominance hierarchies in winter song sparrows. *Condor*, **78**, 567–569.

Krebs, J. R. (1970) Regulation of numbers in the Great Tit (Aves: Passeriformes). *Journal of Zoology, London*, **162**, 317–333.

Krebs, J. R. (1971) Territory and breeding density in the Great Tit, *Parus major* L. *Ecology*, **52**, 1–22.

Krebs, J. R. (1976) Bird song and territory defence. *New Scientist*, **70**, 534–546.

Krebs, J. R. (1977a) Song and territory in the Great Tit. In *Evolutionary Ecology*, ed. Stonehouse, B. and Perrins, C. M. London: Macmillan.

Krebs, J. R. (1977b) The significance of song repertoires: the Beau Geste hypothesis. *Animal Behaviour*, **25**, 575–578.

Lack, D. (1958) A quantitative study of British Tits. *Ardea*, **46**, 91–124.

Lack, D. (1966) *Population studies of birds*. Oxford: Clarendon Press.

Löhrl, H. (1973) Einfluss der Brutraumfläche auf die Gelegegrösse der Kohlmeise (*Parus major*). *Journal für Ornithologie*, **114**, 339–347.

Maynard Smith, J. (1974) *Models in Ecology*. London: Cambridge University Press.

Murdoch, W. W. (1970) Population regulation and population inertia. *Ecology*, **51**, 497–502.

Nilsson, S. G. (1975) Kullstorlek och Lackningsfrangang i holkar och naturliga hal. *Vår Fågalvärld*, **34**, 207–211.

Pennycuick, L. (1969) A computer simulation of the Oxford Great Tit population. *Journal of Theoretical Biology*, **22**, 381–400.

Perrins, C. M. (1965) Population fluctuations and clutch-size in the Great Tit, *Parus major* L. *Journal of Animal Ecology*, **34**, 601–647.

Perrins, C. M. (1966) The effect of beech crops on Great Tit populations and movements. *British Birds*, **59**, 419–432.

Perrins, C. M. and Moss, D. (1975) Reproductive rates in the Great Tit. *Journal of Animal Ecology*, **44**, 695–706.

Risser, A. C. Jr. (1975) Experimental modifications of reproductive performance by density in captive starlings. *Condor*, **77**, 125–132.

Simon, C. A. (1975) The influence of abundance of food on territory size in the ignanid lizard *Scleropus jarrovi*. *Ecology*, **56**, 993–998.

Slaney, P. and Northcote, T. G. (1974) The effects of prey abundance on density and territorial behaviour of young rainbow trout *Salmo gairdneri* in laboratory stream channels. *Journal of the Fisheries Research Board of Canada*, **31**, 1201–1209.

Stenger, J. (1958) Food habits and available food of Ovenbirds in relation to territory size. *Auk*, **75**, 335–346.

Varley, G. C., Gradwell, G. R., and Hassell, M. P. (1973) *Insect population ecology: an analytical approach*. Oxford: Blackwell Scientific.

Watson, A. and Moss, R. (1972) A current model of population dynamics in the red grouse. In *Proceedings of the fifteenth International Ornithology Congress*, pp. 134–149.

Webber, M. I. (1975) Some aspects of the non-breeding population dynamics of the Great Tit (*Parus major*). D.Phil. thesis, University of Oxford.

Residents and transients in wood mouse populations

J. R. FLOWERDEW

Department of Applied Biology, University of Cambridge, Cambridge, England

Introduction

The population dynamics of the wood mouse, *Apodemus sylvaticus*, within woodland habitats are amongst the best documented of those of wild rodents. Since the study of Elton *et al.* (1931), the annual fluctuation in numbers has been described repeatedly from further studies near Oxford and from elsewhere in Britain and Europe (e.g. Evans, 1942; Miller, 1958; Bergstedt, 1965; Crawley, 1970; Southern, 1970a, 1970b; Louarn and Schmitt, 1972; Flowerdew, 1972, 1974).

The annual fluctuation consists of high numbers in late autumn or winter followed by a spring decline and often stationary numbers in summer. The variation in the fluctuation which occurs from year to year has been analysed in detail by Watts (1966, 1969). In general the relative changes in numbers over the year may be described in one of three main ways (figure 1). Numbers may be moderate, low, or intermediate in the summer but the even higher numbers found in winter are less variable; despite the great variation found in the summer numbers, the winter numbers only vary by a factor of about two. There is thus an apparent ceiling in winter numbers but far less certainty over summer numbers; indeed, in some years wood mice almost completely disappear from woodland habitat during the summer (Newson, 1960; Tanton, 1965; Smyth, 1968).

From his analysis of 18 years of wood mouse population studies carried out at Oxford, Watts (1969) concluded that:

1. Overwintering success is strongly correlated with the size of the autumn seed (acorn) crop (figure 2).
2. The decrease in density in the spring is density dependent, being slightly greater following higher initial densities.

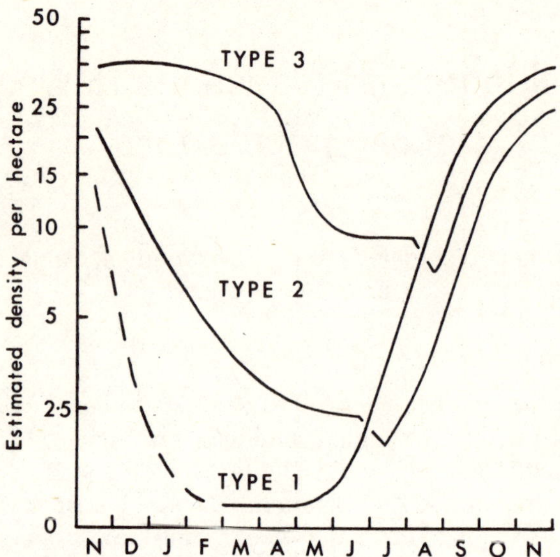

Figure 1 Typical wood mouse population dynamics as suggested by Watts. After Watts (1966).

3. Summer density is stationary, mainly because juveniles survive poorly.
4. Numbers increase in the autumn because survival, especially of juveniles, improves notably. The timing of the increase is density dependent (figure 3) and Watts considered that this was the main factor regulating wood mouse populations. At high density in the summer the time between the start of the increase and the end of the breeding season was shorter, thus setting a limit to winter numbers.

Thus the regulation of the winter numbers seems to be due to a density-dependent process which is initiated in the summer. However, early summer numbers, although determined by the level of the seed crop the previous autumn (figure 2), remain stationary because of poor juvenile survival. A relatively constant level of summer numbers is maintained despite immigration and intensive breeding (Flowerdew, 1974). To explain the processes which limit population numbers we must find a factor which allows the survival of juveniles to change from poor to good and which alters in intensity with summer density.

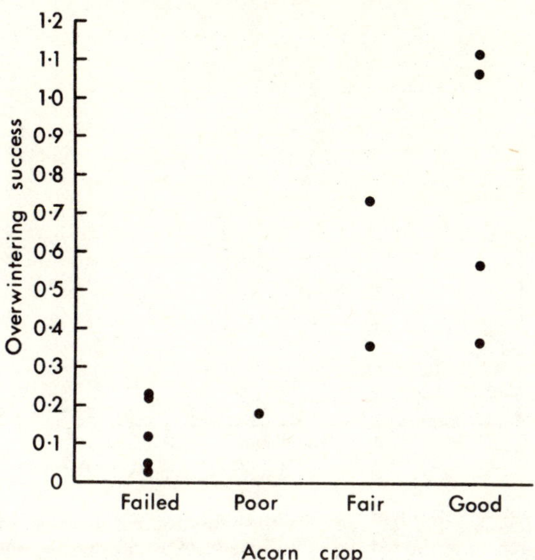

Figure 2 Overwintering success (May–June population less juveniles divided by December population) of wood mice in relation to the acorn crop the previous autumn. After Watts (1969).

Watts (1969) suggested that social behaviour was involved in the changes from high to low and back to high numbers again. The decline in the spring could be the result of increased strife between individuals coincident with the start of the breeding season (February–March). Stationary numbers and poor juvenile survival in summer might be due to antagonism between adults and juveniles which causes a high death rate in the latter. Dispersal was thought not to be important in woodland (Watts, 1970) and the food supply available appeared to bear little relationship to summer density. The increase in numbers in the autumn, the result of increased juvenile survival, could occur by 'interactions between young and old finally turning out in favour of the young' (Chitty and Phipps, 1966). This change in the result of behavioural interactions might be explained partly by the adults becoming less agressive as they went out of breeding condition. However, this could not be the complete answer as breeding sometimes continued into the winter (Smyth, 1966). These ideas involving aggression and social relationships have come from experimental and observational studies of the deer-

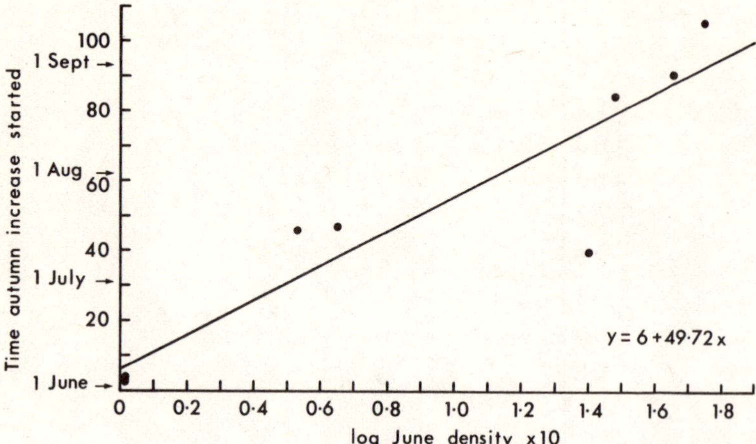

Figure 3 Relationship between the time the autumn increase in wood mouse numbers started and the density of wood mice at the start of the breeding season (June). After Watts (1969).

mouse, *Peromyscus maniculatus*, carried out by Sadleir (1965) and Healey (1967); the population dynamics of the two species are apparently very similar.

These conclusions and speculations based on correlations and comparisons with studies of other, similar, species suggested further questions which could be tested by experimental population and behaviour studies; some of these have been tried (Flowerdew, 1971, 1972, 1974). The experiments, in general, corroborate the conclusions and suggestions of Watts (1969) but there are still gaps in our understanding of *Apodemus* population regulation and change, especially where social behaviour is concerned.

A further question which is not always easy to answer is whether changes in survival are mainly reflections of changes in actual mortality or simply changes in dispersal (emigration), or both. Also, there is now evidence that immigration from one habitat to another can cause the sharp upturn in numbers in the autumn (Corke, 1970, 1974; Leigh Brown, in press) and it is important to place this information in the context of *Apodemus* population dynamics in general.

The fact that some individuals remain in their natal area and others move to new areas allows one to categorize individuals into residents and transients. Transient has, in the past, been defined as an individual which is caught once or perhaps only a few times and this should include those individuals which are making occasional sorties, or dispersing, or which are wanderers (Watts, 1970). I wish to extend this definition to include any individual which is immigrating or emigrating, that is, one which is not a permanent member of a resident population for its whole life.

Movement, rather than mortality or natality, may play an important part in the observed changes in numbers of wood mice; in the following sections I will try to assess the roles of residents and transients in determining the numbers of wood mice present in woodland populations at different times of the year.

Population dynamics, transients and mortality

The winter period of high density

Watts (1969) considers that wood mice reach the limit of their food supply in oak/ash/sycamore woodland one winter in two. This is likely to occur when the mast crop has failed or when it is patchy and it seems even more likely in less productive woodland and perhaps arable land. Circumstantial evidence to show that autumn/winter densities vary in relation to the available food supply comes from Hansson's (1971) index trapping in southern Sweden where a high positive correlation ($p < 0.01$) was found between the catch per 100 trap nights in September–November and the seed production the same autumn.

Experimental studies also implicate the food supply in determining winter density. I added wheat to a woodland study area and compared numbers with a control (Flowerdew, 1972). Wood mice showed higher minimum numbers in one year from February to May and underwent a slower decline in another year (figure 4). Both years showed at least double the proportion of immigrants on the experimental area in late winter.

The question remains 'from where do such immigrants come?'. Evidence from a short experiment in winter (Flowerdew, 1976) seems to indicate that much of the increase in density on an area with additional food comes from mice moving only short

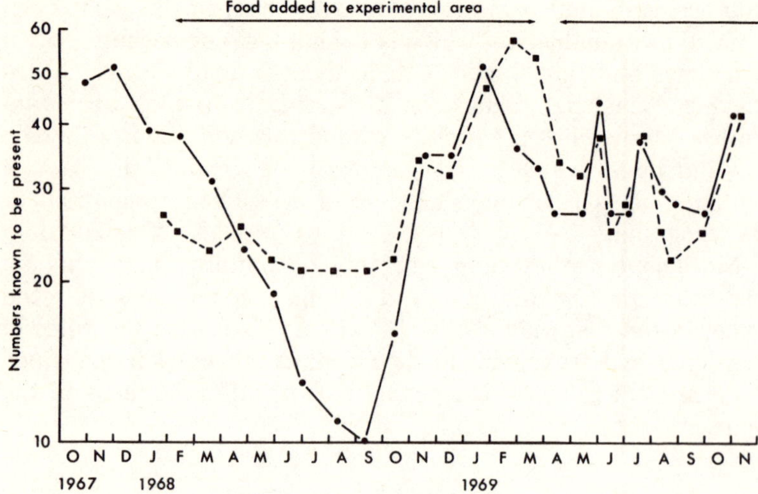

Figure 4 Numbers of wood mice known to be present on an experimental area with additional food (■) and a similar control area (●) in Marley Wood, near Oxford. After Flowerdew (1972).

distances (up to 35 m) (figure 5). Two sides of the experimental area where additional food had been supplied were occupied by control grids and movements into the experimental area were monitored. Only one out of seven recorded moves onto the experimental area was of 75 m or more, the other six mice moved 35 m or less. Thus long-distance movement seems likely to be relatively unimportant in the movements of mice in winter. Trap-revealed movements are usually small during this season (Miller, 1958; Crawley, 1969).

The decline from winter to early summer

Declines in the numbers of small rodents are common demographic events (Chitty and Phipps, 1966) which often occur amidst an apparently abundant food supply (Chitty *et al.*, 1968). However, there is some evidence concerning the involvement of food supply in the decline from the experimental addition and withdrawal of food (figure 4) (Flowerdew, 1972). When food was withdrawn for a few weeks in March and April numbers on the experimental area dropped markedly from the high numbers

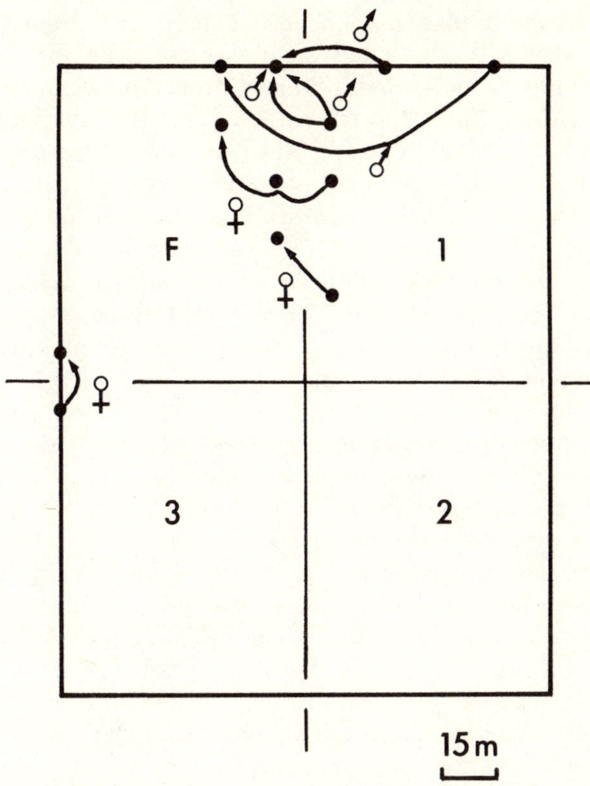

Figure 5 Movements of wood mice from control areas (1–3) to an experimental grid (F) with additional food in Madingley Wood, near Cambridge. Movements took place over a two week period. After Flowerdew (1976).

which had been present earlier in March and February; however, numbers declined still further after the addition of food had recommenced. Further, more general, evidence comes from Watts' (1969) correlation between overwinter survival and the previous autumn's acorn crop (figure 2). Thus the numbers present at the start of the summer are related to the previous food supply, of which some may still be present. Watts (1968, 1969) also gives evidence of abundant insect and herbaceous plants being present during declines so food supply cannot be limiting numbers over all this period even if it does so for part of it.

The possibility that the decline is caused by increased intra-specific strife (Watts, 1969), as it seems to be for deermice (Healey, 1967), has gained support from behavioural studies (Hedges, 1966; Gurnell, 1972) which show that the aggressive behaviour of males towards strange males increases from winter to spring, concurrent with the start of the breeding season. The fact that this increase in aggression cannot always be shown (Flowerdew, 1971, 1974) may be the result of fluctuating aggressive and tolerant phases in wood mouse behaviour or of fluctuating 'social' and 'asocial' phases of activity (Bovet, 1972a, 1972b). Some behavioural information has been obtained from the field. Kikkawa (1964) did not observe aggressive encounters at a baited trap between November and February. However, Brown (1966, 1969) did observe dominant–subordinate aggressive relationships in summer and she states that the more usual evidence of a more dominant animal is that of the submissiveness of other mice in his presence. Overt aggression is perhaps uncommon between established members of a social group; a recent study by Garson (1975) also suggests that subordinates avoid dominants by having different patterns of activity, similar to the phases of behaviour observed by Bovet (1972b). Thus laboratory studies of behaviour have inherent disadvantages and give variable results but field observations may only rarely show the rejection of an individual from a social group.

Dispersal, predation, and starvation are the three most likely causes of the decline from winter to spring. There is no doubt that some dispersal occurs at this time; between March and June 1969 I recorded four long movements from one area of woodland into another. These movements were over 130 m and would therefore be categorized by Watts (1970) as dispersal. Also, movements out of woodland into arable land have been recorded by Bergstedt (1966) and by Corke (1970, 1974). The reasons for the lack of evidence for dispersal within the same woodland may be that dispersing animals are trap-shy whilst in process of moving from one area to another (Watts, 1970); dispersing animals may also be subject to heavy predation as suggested by Healey (1967) or dispersal within the same habitat may really be uncommon.

The effect of predation is difficult to assess. King (personal communication) observed that weasels (*Mustela nivalis*) were apparently not significant predators of wood mice in Marley Wood in spring 1969, but considered that factors peculiar to this year, i.e. the reduction in the number of weasels (King, 1975), probably influenced this result. Although weasel predation did not account for all the observed mortality each month in the wood mouse population the precise effect could not be calculated. The tawny owl (*Strix aluco*) takes more mice during late winter and spring than at other times of the year (Southern, 1954, 1969) but the significance of owl predation has not been determined. Also, behavioural studies are needed to see if any mortality is indirect and socially induced (Errington, 1946) or if it occurs regardless of social position.

Starvation has been studied by Leigh Brown (1976). He found highly significant differences in liver glycogen levels between animals of different phosphoglucomutase genotypes after overnight fasting. This indicates that starvation could affect the quality of the population and also possibly the density at times of food shortage.

Immigration by transients in winter is well documented in population studies of wood mice (Evans, 1942; Flowerdew, 1972). It seems likely that as overwinter survival is so well correlated with the previous acorn crop (Watts, 1969) and as mice may move in response to the available food supply (Flowerdew, 1976), some compensatory movement may quickly make up for losses in numbers due to heavy predation or other mortality factors; Brown (1954) noted rapid immigration into a depopulated area. Despite this possibility, numbers still decline into the spring coincident with some dispersal and increasing male aggression.

Stationary summer numbers and patterns of survival

During the summer months numbers often remain unchanged despite short-lived influxes of adults and juveniles (Flowerdew, 1974) (figure 6). As food supply in the summer is apparently not limiting (Watts, 1969; Flowerdew, 1972) it seems likely that social behaviour is responsible for the poor survival of immigrants and recruits. This idea receives support from studies comparing unmanipulated, control, areas with areas where some or all individuals have been removed by trapping. In a

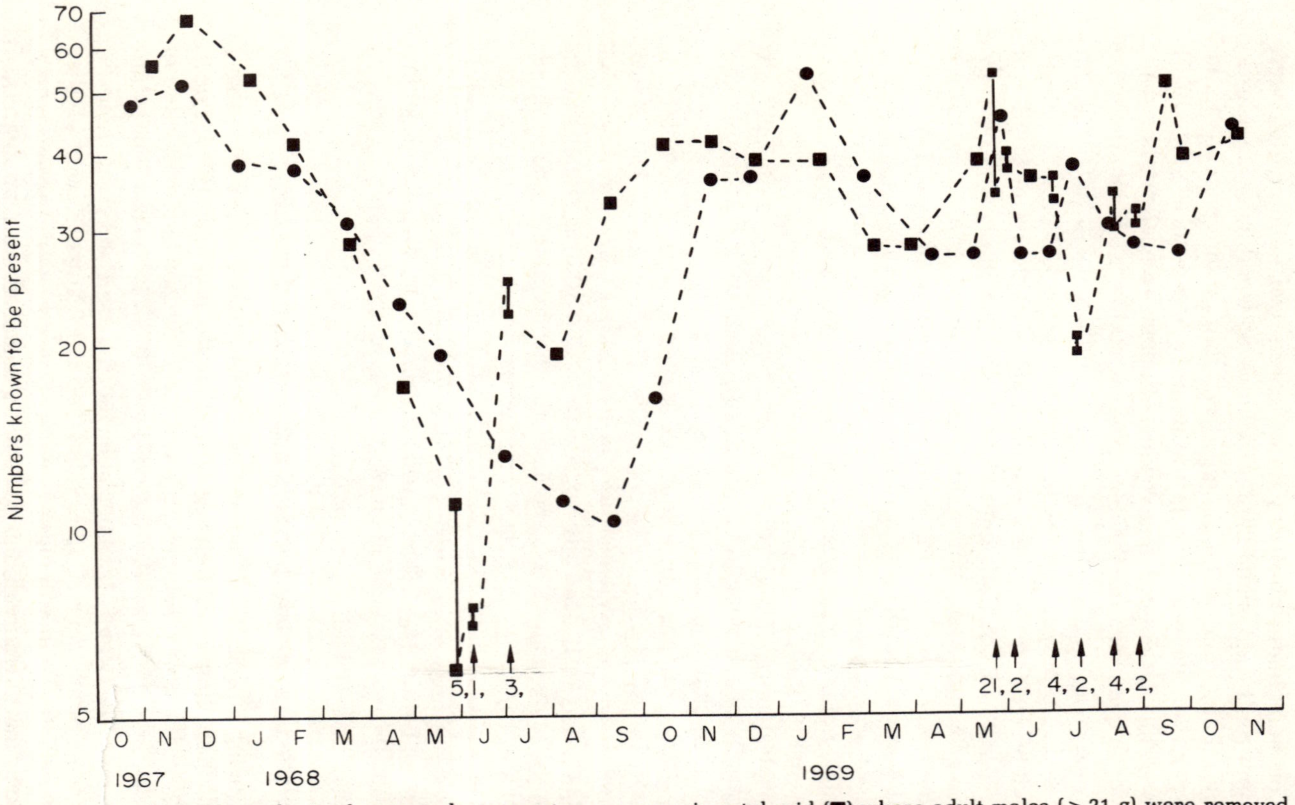

Figure 6 Numbers of wood mice known to be present on an experimental grid (■) where adult males (> 21 g) were removed and on a nearby control area (●). Numbers of males removed at each trapping (——) are shown below the points. From Flowerdew (1974) with permission.

Table 1 Minimum survival of six artificial immigrants on to occupied and trapped-out grids (from Flowerdew, 1974)

Expt	Date	No. surviving to day 7 or longer			No. of residents on occupied grid	
		Trapped-out grid	Occupied grid	Difference	Start	Finish
1	12 June	2	1	+1	11	9
2	28 June	6	3	+3	12	9
3	25 Aug.	5	4	+1	8	7
4	13 Sept.	3	2	+1	6	3
5	25 Sept.	4	1	+3	12	12

number of short-term experiments where all trappable animals were removed, the artificial immigrants placed in the experimental areas survived much better than on the control areas (table 1). Similar long-term experiments where adult (presumably dominant) males were removed (Flowerdew, 1974) provide good evidence that numbers are limited by social factors. In both years of the removal experiment (figure 6) the lack of adult males during the early summer apparently caused an increase in juvenile and subadult survival and an increase in immigration/recruitment. In each year the time of the increase in numbers which usually occurs in the autumn was advanced so that numbers on the experimental area were double those of the control by July in 1968 and nearly double by September 1969.

Laboratory experiments have confirmed that established adults of both sexes, but particularly males, were detrimental to the growth and survival of juveniles (Flowerdew, 1974). Thus, in the wild, juveniles may well be forced to disperse. However, juvenile dispersal is not commonly recorded. Watts (1970) noted only one movement (of 440 m) by an 11 g female. I have noted eight dispersal moves out of woodland (Flowerdew, 1974) and two of these moves, one of 330 m and the other of 140 m, were by 15 g juvenile males. Neither of these juveniles were recaptured after their dispersal movements. From the available evidence it seems that the dispersal of juveniles does occur but that the survival of dispersing animals is poor. Detailed studies of genetics or of radioactively tagged individuals (e.g. Hilborn and Krebs, 1976) will be needed to define the role of dispersal in wood mouse population dynamics.

The autumn increase

The number of juveniles surviving during the autumn is markedly greater than during the summer (Watts, 1969). Juveniles do not make dispersal movements after August (Watts, 1970), or possibly September (Flowerdew, unpublished). Juveniles born late in the year remain close to their place of birth compared with those born earlier in the year (Randolph, 1973) and movements generally are more restricted during the autumn and winter than during the summer. Watts (1969) found it difficult to explain the increased survival of juveniles simply in terms of a change from adult aggression in summer to adult docility and tolerance in autumn and winter in parallel with the end of the breeding season. This was because breeding sometimes continues into the winter (Smyth, 1966). However, there is good evidence to show that the overwintered adults from the previous year are present only in small numbers by the following autumn (figure 7) (Flowerdew, 1972). Thus if adults which have matured during the year of their birth react in a more tolerant way to juveniles than do overwintered adults then this might explain the coincidence of the autumn increase with continued breeding in some years. Chitty and Phipps (1966) and Wilson (1973) suggest that the mortality of overwintered adults is an important prerequisite for the good survival of juvenile voles.

Thus changes in behaviour and changes in population structure are involved in the change in survival which occurs in the autumn. Leigh Brown (in press) found, however, that in some populations immigration appears to be the main factor causing the increase in numbers. His study of a Leicestershire woodland adjacent to a cornfield showed that dramatic increases in the numbers of mice in the woodland occurred after harvesting and many presumed immigrants showed a different genotype for a phosphoglucomutase polymorphism in the blood from that previously found in the woodland population. This genotype (*cc*) was selected against during the winter so that it occurred at a low frequency during most of the following year. Although the sample sizes were too small to prove that immigration was definitely involved in the appearance of the *cc* genotype some movement between habitats was postulated to account for the continued existence of this widespread polymorphism. Also, Corke's (1970, 1974) study of a similar habitat showed that

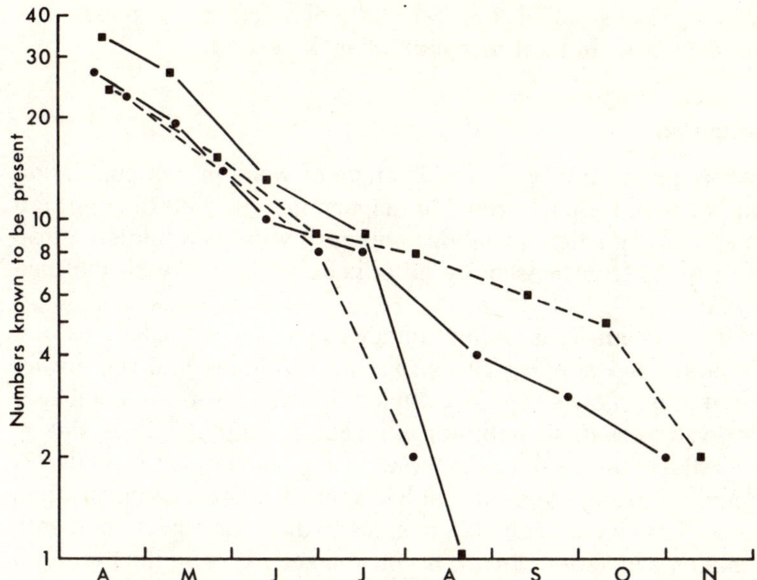

Figure 7 Survivorship curves of cohorts of overwintered wood mice known to be present in April. Experimental area with additional food (■); control area (●); 1968 (— — —); 1969 (———). Both trapping grids were in Marley Wood, near Oxford. After Flowerdew (1972).

individuals returned to the woodland from fields in the autumn and thus contributed to the increase in numbers. In areas of continuous woodland or other suitable habitat the contribution to the autumn increase by immigrants is likely to be small; when unsuitable habitat (pasture) bordered Leigh Brown's woodland grid numbers still increased in the autumn but to a much lower level (*circa* 25 compared with 80). A study of wood mice in arable fields lacking alternative habitats (R. E. Green, personal communication) shows little evidence of immigration in the autumn. There is an upturn in juvenile survival in the autumn similar to that in continuous woodland and changes in survival account for much of the observed changes in numbers. In the field habitats changes in numbers mimic the fluctuations observed in woodland and densities of 15–25/ha have been observed in the autumn/winter peaks. Thus there is no suggestion from this continuous field habitat that it is simply acting as a summer reservoir for a woodland emigrant population as found by Corke (1974). Further study of wood mouse population dynamics over habitat discontinuities is needed.

Discussion

Transients are involved in each stage of wood mouse population dynamics but their probable importance in determining the density on a particular habitat will vary with its situation, close or far away from seasonally suitable habitats, and with the time of year.

Winter transients allow numbers to be adjusted to the available food supply. This probably continues into the spring and at least some of the loss during the winter–spring decline is due to dispersal. In habitats adjacent to suitable fields this is particularly noticeable. Dispersal in the summer by juveniles is known to occur and probably is much more common than researchers have been able to show to date. Immigration of both juveniles and subadults occurs in summer but the effect on total numbers is short-lived (Flowerdew, 1974). In the autumn juveniles cease to disperse and they survive well; in areas adjacent to harvested or other suitable crops immigration will also aid the increase in numbers which occurs between summer and winter.

The evidence for social causes of dispersal or poor survival is mainly indirect. However, interactions between adult males and juveniles are probably the cause of juvenile dispersal and the antagonism of residents towards transients in the summer will allow numbers to be stabilized while enough overwintered adult males remain to maintain the social system. The social grouping present in the summer is thought to be based on a 'super-family' of a small number of adult males and females within the territory of a dominant male (Brown, 1966, 1969); within such a clan pair bonds between males and females of considerable intensity would be found (Garson, 1975). Strangers and most new juveniles would be excluded from the group and forced to disperse. With the onset of autumn most of the overwintered individuals will have died and any newly matured adults will start reproductive regression except in good seed years. The change in social structure is associated with the toleration and non-dispersal of juveniles and the numbers in the population increase to the autumn–winter peak. The social system may well change from a territorial group structure to a simple hierarchical system allowing more flexibility of movement to adjust the population to the food supply available. This change in social system was first proposed by Watts (1966) and although the detail of the changes in social behaviour are still in need of confirmation there is much circumstantial evidence to support it.

The density-dependent nature of the start of the autumn increase fits in well with the postulated role of the change in social system and the observed changes in age structure of the population in autumn. Low numbers of adults in the summer will reach a density where the social system breaks down early. If there are high numbers of adults the subdominant can take over from the dominant and maintain the group territory. Thus numbers in winter are adjusted by this regulating factor to be high enough to provide a viable breeding population for the next year without completely overeating the winter food supply. However, in some years the acorn crop fails and overwinter survival is poor but the nature of the density-dependent control of the start of the increase allows the population to increase at the start of the summer and compensate for the losses which occurred over winter. Because the acorn crop failed, on average, one year in two Watts (1969) predicted that areas with a stable winter food supply would have more regular fluctuations in numbers. I further predict that areas of continuous habitat should show more regular and stable fluctuations than others adjacent to seasonally suitable areas. Transient animals, particularly immigrants, are likely to be important in the dynamics of the wood mouse in the latter areas.

Behavioural interactions seem to play an important part in the population dynamics of the wood mouse and the need for further research has already been stressed. The roles of genetic factors, predation, and starvation should not be overlooked and it is hoped that future research will also pay attention to these aspects of the problem.

Acknowledgements

I wish to thank Dr Andy Leigh Brown, Dr David Corke, Rhys Green, Dr John Gurnell, and Dr C. M. Miller (King) for allowing me access to their unpublished data. I am very grateful to Dr Andy Leigh Brown for many discussions on wood

mouse population dynamics and to Dr H. N. Southern, Dr C. F. Mason, and Dr J. Clevedon Brown for critically reviewing the manuscript.

References

Bergstedt, B. (1965) Distribution, reproduction, growth and dynamics of the rodent species *Clethrionomys glareolus* (Schreber), *Apodemus flavicollis* (Melchior) and *Apodemus sylvaticus* (Linné) in southern Sweden. *Oikos*, **16**, 132–160.

Bergstedt, B. (1966) Home ranges and movements of *Clethrionomys glareolus*. *Apodemus flavicollis* and *A. sylvaticus* in Sweden. *Oikos*, **17**, 150–157.

Bovet, J. (1972a) On the social behaviour in a stable group of Long-tailed field mice (*Apodemus sylvaticus*). I. An interpretation of defensive postures. *Behaviour*, **41**, 43–54.

Bovet, J. (1972b) On the social behaviour in a stable group of Long-tailed field mice (*Apodemus sylvaticus*). II. Its relations with distribution of daily activity. *Behaviour*, **41**, 55–67.

Brown, L. E. (1954) Small mammal populations at Silwood Park Field Centre, Berkshire, England. *Journal of Mammalogy*, **35**, 161–176.

Brown, L. E. (1966) Home range and movement in small mammals. *Symposium of the Zoological Society, London*, **18**, 111–142.

Brown, L. E. (1969) Field experiments on the movements of *Apodemus sylvaticus* L. using trapping and tracking techniques. *Oecologia*, **2**, 198–222.

Chitty, D. and Phipps, E. (1966) Seasonal changes in survival in mixed populations of two species of vole. *Journal of Animal Ecology*, **35**, 313–331.

Chitty, D., Pimentel, D., and Krebs, C. J. (1968) Food supply of overwintered voles. *Journal of Animal Ecology*, **37**, 113–120.

Corke, D. (1970) The local distribution of the yellow-necked mouse (*Apodemus flavicollis*). *Mammal Review*, **1**, 62–66.

Corke, D. (1974) *The comparative ecology of the two British species of the genus* Apodemus *(Rodentia, Muridae)*. Ph.D. thesis, University of London.

Crawley, M. C. (1969) Movements and home ranges of *Clethrionomys glareolus* Schreber and *Apodemus sylvaticus* L. in north-east England. *Oikos*, **20**, 310–319.

Crawley, M. C. (1970) Some population dynamics of the Bank vole, *Clethrionomys glareolus* and the wood mouse, *Apodemus sylvaticus* in mixed woodland. *Journal of Zoology, London*, **160**, 71–89.

Elton, C. S., Ford, E. B., Baker, J. R., and Gardner, A. D. (1931) The health and parasites of a wild mouse population. *Proceedings of the Zoological Society of London*, 657–721.

Errington, P. L. (1946) Predation and vertebrate populations. *Quarterly Review of Biology*, **21**, 144–177, 221–245.

Evans, F. C. (1942) Studies of a small mammal population in Bagley Wood, Berkshire. *Journal of Animal Ecology*, **11**, 182–197.

Flowerdew, J. R. (1971) *Population regulation of small rodents in relation to social behaviour and environmental resources*. D.Phil. thesis, University of Oxford.

Flowerdew, J. R. (1972) The effect of supplementary food on a population of wood mice (*Apodemus sylvaticus*). *Journal of Animal Ecology*, **41**, 553–566.

Flowerdew, J. R. (1974) Field and laboratory experiments on the social behaviour and population dynamics of the wood mouse (*Apodemus sylvaticus*). *Journal of Animal Ecology*, **43**, 499–511.

Flowerdew, J. R. (1976) The effect of a local increase in food supply on the distribution of woodland mice and voles. *Journal of Zoology*, London, **180**, 509–513.

Garson, P. J. (1975) Social interactions of woodmice (*Apodemus sylvaticus*) studied by direct observation in the wild. *Journal of Zoology, London*, **177**, 496–500.

Gurnell, J. (1972) *Studies on the behaviour of wild wood mice*, Apodemus sylvaticus. Ph.D. thesis, University of Exeter.

Hansson, L. (1971) Small rodent food, feeding and population dynamics. *Oikos*, **22**, 183–198.

Healey, M. C. (1967) Aggression and self-regulation of population size in deermice, *Ecology*, **48**, 377–392.

Hedges, S. R. (1966) *Studies on the behaviour, taxonomy and ecology of* Apodemus sylvaticus *(L.) and* Apodemus flavicollis *(Melchior)*. Ph.D. Thesis, University of Southampton.

Hilborn, R. and Krebs, C. J. (1976) Fates of disappearing individuals in fluctuating populations of *Microtus townsendii*. *Canadian Journal of Zoology*, **54**, 1507-1518.

Kikkawa, J. (1964) Movement, activity and distribution of the small rodents *Clethrionomys glareolus and Apodemus sylvaticus* in woodland. *Journal of Animal Ecology*, **33**, 259–299.

King, C. M. (1975) The home range of the weasel (*Mustela nivalis*) in an English woodland. *Journal of Animal Ecology*, **44**, 639–668.

Leigh Brown, A. J. (1976) Ecological genetics in *Apodemus sylvaticus*. Ph.D. thesis, University of Leicester.

Leigh Brown, A. J. (*In press*) Genetic changes in a population of fieldmice (*Apodemus sylvaticus*) during one winter. *Journal of Zoology, London*.

Louarn, H. Le and Schmitt, A. (1972) Relations observées entre la production de faines et la dynamique de population du mulot, *Apodemus sylvaticus* L. en forêt de Fontainebleau. *Annales des Sciences forestières*, **30**, 205–214.

Miller, R. S. (1958) A study of a wood mouse population in Wytham Woods, Berkshire. *Journal of Mammalogy*, **39**, 477–493.

Newson, R. (1960) *The ecology of vole and mouse populations in different habitats*. D.Phil. thesis, University of Oxford.

Randolph, S. E. (1973) A tracking technique for comparing individual home ranges of small mammals. *Journal of Zoology, London*, **170**, 509–520.

Sadleir, R. M. F. S. (1965) The relationship between agonistic behaviour and population changes in the deermouse, *Peromyscus maniculatus* (Wagner). *Journal of Animal Ecology*, **34**, 331–352.

Smyth, M. (1966) Winter breeding in woodland mice, *Apodemus sylvaticus*, and voles, *Clethrionomys glareolus* and *Microtus agrestis*, near Oxford. *Journal of Animal Ecology*, **35**, 471–485.

Smyth, M. (1968) The effects of the removal of individuals from a population of bank voles *Clethrionomys glareolus*. *Journal of Animal Ecology*, **37**, 167–183.

Southern, H. N. (1954) Tawny owls and their prey. *Ibis*, **96**, 384–410.

Southern, H. N. (1969) Prey taken by tawny owls during the breeding season. *Ibis*, **111**, 293–299.

Southern, H. N. (1970a) Ecology at the cross-roads. *Journal of Animal Ecology*, **39**, 1–11.

Southern, H. N. (1970b) The natural control of a population of tawny owls (*Strix aluco*). *Journal of Zoology, London*, **162**, 197–285.

Tanton, M. T. (1965) Problems of live-trapping and population estimation for the wood mouse *Apodemus sylvaticus* (L). *Journal of Animal Ecology*, **34**, 1–22.

Watts, C. H. S. (1966) *The ecology of woodland voles and mice with special reference to movement and population structure.* D.Phil. thesis, University of Oxford.

Watts, C. H. S. (1968) The foods eaten by wood mice, *Apodemus sylvaticus*, and bank voles, *Clethrionomys glareolus*, in Wytham Woods, Berkshire. *Journal of Animal Ecology*, **37**, 25–41.

Watts, C. H. S. (1969) The regulation of wood mouse (*Apodemus sylvaticus*) numbers in Wytham Woods, Berkshire. *Journal of Animal Ecology*, **38**, 285–304.

Watts, C. H. S. (1970) Long distance movement of bank voles and wood mice. *Journal of Zoology, London*, **161**, 247–256.

Wilson, S. (1973) The development of social behaviour in the vole (*Microtus* agrestis). *Zoological Journal of the Linnean Society*, **52**, 45–62.

Lemur behaviour and populations

ALISON JOLLY

School of Biological Sciences, University of Sussex, Falmer, Brighton, England

Since the publication of Wynne-Edwards' *Animal Dispersion in Relation to Social Behaviour* (1962), many of his principles have become widely accepted. The initial furor, and the arguments which tended to polarize into all-or-nothing positions, have turned instead into programmes of research. A similar furor now surrounds E. O. Wilson's *Sociobiology* (1975), which amplifies many of Wynne-Edwards' earlier ideas.

Primatologists take for granted that dominance hierarchies and territoriality regulate animals' access to key resources. We even have a few empirical studies which correlate dominance with reproductive success. The interesting questions now seem to be: in what species, populations, or circumstances does an individual maximize its fitness by a 'selfish' personal success; in what circumstances does 'altruistic' kin selection outweigh immediate personal advantage; and when does conservation, or destruction of the environment, impinge as a selective force on the individual and its kin group. By recognizing that all three aspects reflect back on individual fitness, it is possible to ask questions about social behaviour in terms of relative selective pressures on an individual in differing circumstances or at different stages of its life cycle.

At first sight, the primates seem to be the ideal mammalian group for the study of social behaviour and population structure. Their personal relationships are elaborate, obvious, and easily understood. They are easy to understand because the primatologist shares primate sensory equipment and social perceptions. Besides, primates, compared to other mammals of similar body size, are almost invariably longer-lived, slower breeding, K-selected. A monkey which may live for twenty years or more, must encounter most local environmental fluctuations during its lifetime.

The problem is that most primate's lifetimes are too long. We have cross-sectional age structures, but pathetically little longitudinal fieldwork has been carried out. Some artificially provisioned groups have been studied for ten years or more: the Gombe Stream chimpanzees, Japanese macaques, and rhesus monkeys in artificial colonies off Puerto Rico. A few populations with natural ranges have been recensused after five years or more: Amboseli and Queen Elizabeth Park baboons, Amboseli vervets, Barro Colorado howlers, Abu Langurs, one group of Malaysian siamang. Nowhere, as yet, do we have data on individual's reproductive success in an undisturbed environment.

Malagasy lemurs

Malagasy lemurs provide a cross-check on observations of other primates, since they are a primate branch which has been isolated since the Eocene. They are slow breeding, but not so slow as many monkeys. The 60 g mouse-lemur (*Microcebus murinus*), for instance, mates at 8½ months, has a 3½ month gestation period and produces one, or occasionally two, offspring a year. The 2.7 kg *Lemur catta*, the ringtail, and 3.6 kg sifaka, *Propithecus verreauxi*, first mate at 2½ years of age, and have a 4½ month gestation period. Half to three quarters of adult females have an infant in any one year. *Indri*, largest of the extant lemurs at 12.5 kg, apparently have infants at three-year intervals. Lemurs, therefore, give some hope to the primatologist who is interested in populations.

In what ways do lemur social structures resemble those of 'higher' primates (table 1)? Of some twenty lemur species, seven or eight forage alone, unlike higher primates except for night monkeys and the orang-utans. A surprising four or perhaps five species live as pairs with their offspring, a pattern less common in monkeys and apes, though present in gibbon, siamang, and titi. None seem to have one-male harems, unless *Hapalemur* proves to be harem-forming—a surprising lack. Finally, six or more species live as multi-male troops, at first sight much like the troops of many monkeys.

What may turn out to be the greatest difference is the relations between the sexes. Among all the group-living lemurs which have been studied intensively, *Indri*, *Propithecus verreauxi*, *Lemur fulvus*, and *Lemur catta*, females have priority of access to food and feeding sites. Females can, and do, cuff and supplant males, even the most dominant male of the troop. In short, females dominate males in most senses and situations (Jolly, 1966; Pollock, 1975a; Richard, in press).

This dominance is not the same as being aggressive. *Lemur catta* and *Propithecus* males have more aggressive encounters than females, and thus a more obvious hierarchy, not to say bullying, among themselves. Pair-living *Indri* males play the most active role in challenging neighbouring groups, leaping and wailing at each other. They are just not dominant over their own females.

A second difference is that *Propithecus* males may change troops during the breeding season (Richard, in press), and a quarter of all *Lemur catta* males have been seen to change troops just after the birth season (Budnitz and Dainis, 1975). *Propithecus* troops have also been seen to break up and reassort with neighbours (Jolly, 1966) or to indulge in chain migration of both males and females from one group to the next (Gustafson, in preparation). The state of our data is such that we know migration occurs, but not how much. It is clear, at least, that the troops are not simple kin groups, and that the year-round male hierarchy is not related to mating success in any simple way. Again we need longitudinal data to understand a male's options for reproductive success in his present troop, and his chances of success in his neighbour's.

Population dynamics at Berenty

There is one obvious site for the more intensive study of lemur populations: the Berenty Reserve, established and maintained by M. Henry de Heaulme and his family. It consists of two halves only one of which is protected. The main half of the reserve consists of some 94 hectares of woodlands, with a peninsula of 6–10 ha of euphorbia bush. The reserve lies on an outside curve of the Mandrare River. Beside the river grows lush forest dominated by *Tamarindus indica* trees. Further inland the gallery forest gives way to drier, more open scrub with a discontinuous tree canopy. The reserve itself, in other

Table 1 Comparative data on Malagasy lemurs

Species	Field study	feeding group size range	feeding group size mean	sex ratio: females/males	multi or single male feeding group	home range (km^2)	day range (km)	home range overlap	boundary defence seen
Microcebus murinus	Martin, 1972, 1974	1	1	3		0.002		small	
Microcebus coquereli	Petter *et al.*, 1971, 1975	1–2	1						
Cheirogaleus medius	Petter, 1962	1	1						
Cheirogaleus major	Petter-Rousseaux, 1962	1	1						
Phaner furcifer	Petter *et al.*, 1971; 1975	1–2	1						
Lepilemur mustelinus	Charles-Dominique and Hladik, 1971, 1974 Sussman, 1972 Petter *et al.*, 1971, 1975	1	1			0.002	0.27	large	yes
Hapalemur griseus	Petter and Peyrieras, 1970, 1975	3–5							
	Pollock, in press	3–6	3						
Hapalemur simus	Petter and Peyrieras, 1975								
Lemur variegatus	Petter, 1962	2–4	3						
	Pollock, in press	2–4	3						

Lemur catta	Jolly, 1966	2–24	17	1.0	MM	0.06		small	no
	Jolly, 1972b	8–12	12	0.9	MM			large	no
	Sussman, 1972, 1974	5–20	18	1.1	MM	0.7	0.9	large	no
	Budnitz and Dainis, 1975	5–20	13	1.2	MM	0.06–0.23	0.75		no
	Mertl, in preparation	8–21	13		MM			variable	no
Lemur macaco	Petter, 1962	4–15	10						
Lemur fulvus fulvus	Petter, 1962	4–17	9	1.3	MM				
	Pollock, in press	6–8							
	Harrington, 1975	12	12	1.0	MM	0.07	0.12		yes
Lemur fulvus rufus	Sussman, 1972, 1974, 1975	4–17	9	1.3	MM	0.01	0.15		no
Lemur mongoz mongoz	Petter, 1962	6–8							
	Tattersall and Sussman, 1975	2–4	3	1.0	SM	0.01	0.61		
Lemur rubriventer	Petter, 1962		5						
	Pollock, in press	2–4							
Propithecus verreauxi verreauxi	Jolly, 1966, 1972	2–8	4.6	0.7	MM	0.2		small	yes
	Richard, in press	3–7	5.3	1.0	MM	0.5	0.85	small	yes
	Gustafson, in preparation	4–9	6.3	1.2	MM			small	yes
	Richard, in press	4–8	5.1	1.4	MM	6.8–8.5	0.9	small	yes
Propithecus verreauxi coquereli	Richard, in press	4–10	5.0	1.3	MM	6.8–8.5	0.8	large	no
Propithecus diadema diadema	Pollock, in press	2–5							
Avahi laniger	Pollock, in press	2–4							
Indri indri	Pollock, 1975, in press	2–5	3	1	SM	0.23			
Daubentonia madagascariensis	Petter and Peyrieras, 1970b	1							

Notes:
Field data on group size, activity patterns, and population density is largely from Pollock (in press), amplified by checking original authors cited. See Pollock (in press) for more detailed discussion of lemur ranging patterns.

Table 2 Lemur studies at Berenty

Year	Author	Duration J	F	M	A	M	J	J	A	S	O	N	D
1963	Jolly		—	—	—	—	—	—	—	—			
1964	Jolly			—	—								
1969	Klopfer	—											
1970	Hladik, Charles-Dominique									—	—		
	Pariente									—			
	Jolly									—	—		
	Sussman, Schilling											—	—
1971	Richard, Struhsaker									—			
1972	Budnitz, Dainis				—	—	—	—	—	—	—	—	—
1973	Budnitz, Dainis	—	—	—	—	—							
1974	Russell, McGeorge						—	—	—	—	—	—	—
	Richard					—							
1975	Mertl, Gustafson		—	—	—	—	—	—					
	Jolly, Ramanantsoa									—			

words, is almost an 'experimental' design of two comparable habitats.

Some sixteen primatologists have studied the lemurs of Berenty during the years from 1963 to 1975 (table 2). Each team 'discovered' that the de Heaulme properties are the best places in Madagascar for detailed studies of undisturbed animals. The studies had various goals, from pilot surveys to concentrated analyses of olfaction, vocalization, or influence of ambient light. It is, therefore, largely accidental that any data can be compared from study to study. On the other hand, the close correspondence between results of people who did not know each other's methods or, usually, each other's results, greatly strengthens the conclusions.

Propithecus verreauxi

The white sifaka, *Propithecus verreauxi*, seems an ideal mammal for the study of individual, kin, and group selection. Individuals are recognizable by their facial patterns. Groups, averaging about five animals, persist from year to year, but they are not simple kin groups, for animals reassort by inter-group migration. With longitudinal data, it would be possible to measure reproductive succes for groups of varying composition. The resources used by each group are delimited by the animals themselves, at least at Berenty, as a mosaic of defended territories.

Above all, the number of adults and subadults seems to have been roughly constant over twelve years, a significant length of time even in terms of this slow-breeding, K-selected mammal (table 3; figure 1). The number of infants and juveniles apparently fluctuated wildly over the same period, but this is due to censusing at different times of year. Infants are born during two weeks in July, arbitrarily become juveniles on the following 1 January, and equally arbitrarily turn into subadults the next 1 January at the age of 1½ years. Young adults make their first attempt to breed at 2½ years, though the males, at least, may not succeed.

The fertility rate is 0.4 to 0.6 infants per female, for censuses in July–August after the birth season. That is, some twenty infants are born to an adult and subadult population of about eighty animals. We have no data on adult recruitment, though infant and juvenile mortality presumably offsets most of the annual increase.

Absolute out-migration from the reserve is probably minor. Sifaka can live in the narrow corridor of euphorbia scrub, but out-migration is probably filtered through other troops in the same fashion as migration within the reserve.

We have two cases of troop reassortment recorded by Jolly (1966) and Gustafson (in press), which involved 'chain migration' of sifaka from one troop to the next. In Gustafson's case seven males, three females, and a juvenile shifted from one troop to another within two weeks. This was not mass exodus, but a kind of relay, with one animal leaving when another arrived. The result shows on figure 1 as the high density population of the four best-known groups in early 1975, reduced to normal by late 1975.

Propithecus verreauxi groups at Berenty maintain a classical mosaic of territories. Although there is overlap of troops ranging into each other's areas, the defended boundaries in practice can be identified down to a single tree. Many of these boundaries are 'natural' ones, at places where trees bridge one of the footpaths, offering a crossing point where a group may watch up and down the trail for intruders. Other, apparently equally suitable bridges, fall within territories, however, so a measure of tradition probably helps to determine the frontiers.

Jolly (1972b) remarked on the constancy of *Propithecus* core

Table 3 *Propithecus verreauxi* age structure

Age	0 year												1 year												2 years								Total adults and subadults	Total adults, subadults, and juveniles
	J	A	S	O	N	D	J	F	M	A	M	J	J	A	S	O	N	D	J	F	M	A	M	J	J	A	S	O	N	D	J	F		
	birth ←				infant		←					juvenile						→	←				subadult						→ breeding adult					
Cohorts																																		
1963–4	7 →									5			8 →									7												
1970–1			7 →						3																									
1975	4 →		3								7 →		6 →		3																			
Censuses: total population																																		
1974 17 g											11																						76	87
1975 17 g											20																						76	96
1975 18 g	19±2													20																			76	96
Censuses: 10 ha																																		
1963 9 g	7													8																			34	42
1964 9 g									5																								41	46
1970 9 g			10													7																	39	46
1971 9 g			10												6																		47	53
1975 8 g											11																						47	58
1975 8 g		10±2										12																					46	58
Totals for 4 troops																																		
A = BN, B = DE																																		
C = WP, E = ST																																		
1963	3											4																					17	21
1964											2											2–4											19	21
1970		5																															20	25
1971		4													2																		18	24
1975											7																						23	30
1975	4											6																					23	32
1975		3													3																		19	25

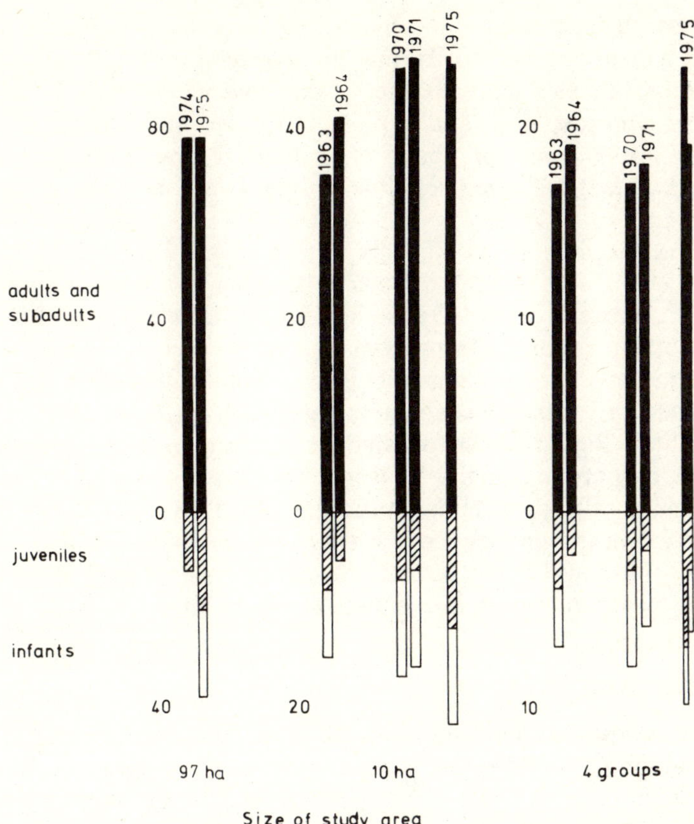

Figure 1 Age distribution for populations of *Propithecus verreauxi*.

areas. In 1975 it was clear that not only most core areas but many sections of frontier across trails were the same as those of 1963. Of course, there were also changes in ranges seen by Jolly in 1963–4 and Gustafson in 1975—the system is not fossilized. However, one of the changes following the 1975 chain migration was the establishment of a new troop, one whose boundaries followed a boundary line which had existed in 1963–4 (Jolly *et al.*, in preparation).

The conclusion seems to be that *Propithecus* at Berenty are exceedingly conservative, and that this is related to the very small scale of their ranges. Their group territories within the rich gallery forest average about two hectares, varying little or not at all from season to season and year to year. All parts of the range are visited during an average fortnight. In these circumstances the possibilities of the environment must be fairly thoroughly explored and exploited. At Berenty, one can come back after years of absence to find groups of *Propithecus* carrying out the same activities on the same branches as when one left.

The *Propithecus* at Barenty are in sharp contrast to those studied by Alison Richard some 30 miles away, in euphorbia scrub or 'spiny desert'. These had larger home ranges, of about 7.0–8.5 ha, with little overlap, but very rare territorial confrontations. They presumably marked their boundaries by scent instead of by the Berenty-style 'arboreal chess games' at the frontier. Richard also studied *P. v. coquereli* in Northern Madagascar, in a highly seasonal deciduous forest. She found almost totally overlapping ranges, with occasional troop confrontation, but no defence of recognizable boundaries. Richard's work thus reveals the Berenty *Propithecus* territorial system to be a local adaptation to a dense, stable habitat.

Lemur catta

Lemur catta at Berenty seem to have an even more stable population than *Propithecus verreauxi* (table 4; figure 2). Births per female in 1970 were 0.8 per cent and infant mortality was about 50 per cent.

Lemur catta's use of the environment varies from year to year and from season to season. Home ranges and day range expand prior to the breeding season, and expand and fragment prior to the male migration. Ranges are virtually without overlap at some periods and almost wholly overlapping at others. Intertroop behaviour varies from mutual feeling to confrontation with screams, spats, cuffing, and biting, largely led by the females.

This variety of behaviour probably corresponds to *Lemur catta*'s niche as a fairly opportunistic animal (for a lemur). It spends 15–65 per cent of its time on the ground in Berenty and 30 per cent at Antseranomby in Western Madagascar (Budnitz and Dainis, 1975; Sussman, 1974). It is the only present-day lemur to range far into scrub and over bare rock faces. It eats fruit as well as leaves, and, alone among the diurnal lemurs,

Table 4 *Lemur catta* age structure

Age	**0 year**												**1 year**												**2 years**								Total adults and subadults	Total adults, subadults and juveniles
	S	O	N	D	J	F	M	A	M	J	J	A	S	O	N	D	J	F	M	A	M	J	J	A	S	O	N	D	J	F	M	A		
	birth infant ←											juvenile				→	←						subadult					→	breeding adult					
Cohorts																																		
1963–4													6 →							5														
1972–3	47 →							28					30 →							14														
Total population censuses																																		
1972	47												30																				123	153
1973									28											14									110				124	152
1975											29																						126	155
Population in 10 ha																																		
1963														6													3		23				26	35
1964									12											5									27				32	44
1970		11												8																			30	38
1972														5																			27	32
1973									6											4									25				29	39
1975												9																					29	38

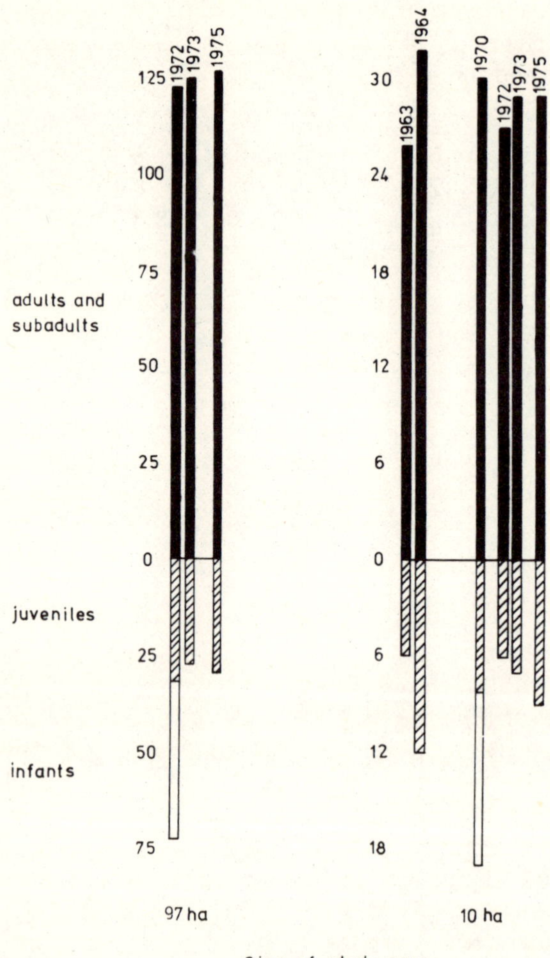

Figure 2 Age distribution for populations of *Lemur catta*.

has been seen to eat insects—the migratory locusts (Budnitz and Dainis, 1975). The ability of some animals to change from mutually exclusive home ranges to a 'time sharing' of key resources is presumably an adaptation to make the most of an environment which is patchy in space and time.

Lemur catta at Berenty are effectively isolated, since they do not live in the euphorbia bush which is the only forest corridor.

They might, conceivably, make an overland progression north along the river, if desperate, but the stable population is probably closed and self-regulating.

Berenty as an island

The habitat at Berenty is not apparently being degraded by the browsing of lemurs but, if anything, it is growing richer as it recovers from human exploitation forty years ago. However, no systematic studies of the plant production have yet been done.

If superficial impressions are right, that the forest is in equilibrium and the lemur populations constant, this seems to be one of the few small, isolated reserves where herbivorous mammal populations are in ecological balance. It is therefore an ideal 'model' for study of this phenomenon, with two halves, two habitats, and two herbivorous species of widely different social structure.

Summary and conclusion

Population data on lemurs is at the stage where Wynne-Edwards began in formulating his hypotheses. We have cross-sectional descriptions which show the variety and interest of lemurs as a primate lineage. Female dominance and inter-group migration raise new questions about the role of the group in lemur ecology. More or less by accident we have longitudinal data for Berenty Reserve, which show that two lemur species have maintained constant populations over twelve years in an 'island' reserve. One species, *Propithecus verreauxi*, here maintains year-round territories; the other, *Lemur catta*, fluctuates widely in its use of the environment. This seems a model site for future critical study of social behaviour and population regulation—assuming that people can maintain enough of the lemurs' habitat to assure any future at all for lemur populations and species.

References

Budnitz, N. and Dainis, K. (1975) *Lemur catta*, ecology and behavior. In *Lemur Biology*, ed. Tattersall, I. and Sussman, R. W. pp. 219–236. New York: Plenum Press.

Charles-Dominique, P. (in press) Observations on *Phaner furcifer*. Paper presented at 6th International Congress of Primatology, Cambridge 1976.

Charles-Dominique, P. and Hladik, C. M. (1971) Le Lépilemur du Sud de Madagascar: Ecologie, alimentation et vie sociale. *Terre et Vie*, **25**, 3–66.

Clutton-Brock, T. H. and Harvey, P. H. (in press) Functional aspects of species differences in feeding and ranging behavior in primates. In *Primate Ecology: Studies of Feeding and Ranging Behavior in Lemurs, Monkeys and Apes*, ed. Clutton-Brock, T. H. London: Academic Press.

Clutton-Brock, T. H. and Harvey, P. H. (in press) Primate ecology and social organization. *Journal of Zoology*, London.

Gustafson, H. (in preparation) *Propithecus verreauxi* groups and territories at Berenty, 1976.

Harrington, J. E. (1975) Field observations of social behavior of *Lemur fulvus fulvus* E. Geoffroy 1812. In *Lemur Biology*, ed. Tattersall, I. and Sussman, R. W. pp. 259–280. New York: Plenum.

Hladik, C. M. and Charles-Dominique, P. (1974) The behavior and ecology of the sportive lemur (*Lepilemur mustelinus*) in relation to its dietary peculiarities. In *Prosimian Biology*, ed. Doyle, G. A. and Walker, A. C. pp. 23–38. London: Duckworth.

Jolly, A. (1966) *Lemur Behavior*. Chicago: University of Chicago Press.

Jolly, A. (1972a) *The Evolution of Primate Behavior*. New York: Macmillan.

Jolly, A. (1972b) Troop continuity and troop spacing in *Propithecus verreauxi* and *Lemur catta* at Barenty (Madagascar). *Folia Primatologia*, **17**, 335–362.

Jolly, A., Gustafson, H., Mertl, A., and Ramanantsoa, G. (in preparation) Population, espace vital, et composition des groupes chez le maki (*Lemur catta*) et le sifaka (*Propithecus verreauxi verreauxi*) à Berenty, République Malagasy. *Bulletin de l'Académie Malgache*.

Klopfer, P. H. and Jolly, A. (1970) The stability of territorial boundaries in a lemur troop. *Folia Primatologia*, **12**, 199–208.

Martin, R. D. (1972) A preliminary field-study of the lesser mouse lemur (*Microcebus murinus* J. F. Miller 1777). *Zeitschrift für Tierpsycholoogie*, Supplement, **9**, 43–89.

Martin, R. D. (1973) A review of the behaviour and ecology of the lesser mouse lemur (*Microcebus murinus* J. F. Miller 1777). In *Comparative Ecology and Behaviour of Primates*, ed. Michael, R. P. and Crook, J. H. pp. 1–68. London: Academic Press.

Mertl, A. S. (1975) Discrimination of individuals by scent in a primate. *Behavioral Biology*, **14**, 505–509.

Mertl, A. S. (1976) Olfactory and visual cues in social interactions of *Lemur catta*. *Folia Primatologia*, 26, 151–161.

Mertl, A. S. (in preparation) Olfactory communication in wild *Lemur catta*.

Pariente, G. (1974) Influence of light on the activity rhythms of two Malagasy lemurs: *Phaner furcifer* and *Lepilemur mustelinus leucopus*. In *Prosimian Biology*, ed. Martin, R. D., Doyle, G. A., and Walker, A. C. pp. 183–198. London: Duckworth.

Petter, J. J. (1962) Recherches sur l'écologie et l'ethologie des Lémuriens malgaches. *Mémoires du Museum National de l'Histoire Naturelle*, Ser. A, **27**, 1–146.

Petter, J. J. and Peyrieras, A. (1970a) Observations éco-éthologiques sur les lémuriens malgaches du genre Hapalémur. *Terre et Vie*, **17**, 356–383.

Petter, J. J. and Peyrieras, A. (1970b) Nouvelle contributions à l'étude d'un lémurien malgache, le aye-aye (*Daubentonia madagascariensis* E. Geoffrey). *Mammalia*, **34**, 167–193.

Petter, J. J. and Peyrieras, A. (1975) Preliminary notes on the behavior and ecology of *Hapalemur griseus*. In *Lemur Biology*, ed. Tattersall, I. and Sussman, R. W. pp. 281-286. New York: Plenum.

Petter, J. J., Schilling, A., and Pariente, G. (1971) Observations éco-éthologiques sur deux lémuriens malgaches nocturnes: *Phaner furcifer* et *Microcebus coquereli*. *Terre et Vie*, **25**, 287–327.

Petter, J. J., Schilling, A., and Pariente, G. (1975) Observations on the behaviour and ecology of *Phaner furcifer*. In *Lemur Biology*, ed. Tattersall, I. and Sussman, R. W. pp. 209–218. New York: Plenum Press.

Petter-Rousseaux, A. (1962) Recherches sur la biologie de la réproduction des primates inférieurs. *Mammalia*, **26**, Supplement 1, 1–88.

Pollock, J. I. (1975a) The social behaviour and ecology of *Indri indri*. Thesis, University of London.

Pollock, J. I. (1975b) *Indri indri:* A preliminary report. In *Lemur Biology*, ed. Tattersall, I. and Sussman, R. W. pp. 287–312. New York: Plenum Press.

Pollock, J. I. (in press) Spatial distribution and ranging behaviour in lemurs. In Doyle, G. A., ed., in press.

Richard, A. F. (1974b) Intra-specific variation in the social organization and ecology of *Propithecus verreauxi*. *Folia Primatologia*, **22**, 178–207.

Richard, A. F. (1976) Preliminary observation on the birth and development of *P. verreauxi* to the age of 6 months. *Primates*, **17**, 357–366.

Richard, A. F. (in press) *Behavioral Variations: Case Study of a Malagasy Lemur*. Bucknell University Press.

Richard, A. F. and Sussman, R. W. (1975) Future of the Malagasy lemurs: conservation or extinction? in *Lemur Biology*, ed. Tattersall, I. and Sussman, R. W. pp. 335–350. New York: Plenum Press.

Sussman, R. W. (1972) An ecological study of two Madagascan primates: *Lemur fulvus* (Audebert) and *Lemur catta* (Linnaeus). Ph.D. thesis, Duke University.

Sussman, R. W. (1974) Ecological distinctions in sympatric species of *Lemur*. In *Prosimian Biology*, ed. Doyle, G. A. and Walker, A. C. pp. 75–108. London: Duckworth.

Sussman, R. W. (1975) A preliminary study of the behavior and ecology of *Lemur fulvus rufus* Audebert 1800. In *Lemur Biology*, ed. Tattersall, I. and Sussman, R. W. pp. 237–258. New York: Plenum Press.

Sussman, R. W. and Tattersall, I. (1976) Cycles of activity, group composition and diet of *Lemur mongoz mongoz* Linnaeus 1766 in Madagascar. *Folia Primatologia*, **26**, 270–284.

Tattersall, I. and Sussman, R. W. (1975) Observations on the ecology and behavior of the lesser mongoose lemur *Lemur mongoz mongoz* Linnaeus (Primates, Lemuriformes) at Ampijoroa, Madagascar. *Anthrop. Papers of the American Museum of Natural History*, **52**, 195–216.

Wilson, E. O. (1975) *Sociobiology: The New Synthesis*. Cambridge, Mass.: Harvard University Press, Belknap.

Wynne-Edwards, V. C. (1962) *Animal Dispersion in Relation to Social Behaviour*. Edinburgh and London: Oliver and Boyd.

The adaptive significance of baboon and macaque social behaviour

JOHN M. DEAG

University of Edinburgh, Department of Zoology, West Mains Road, Edinburgh, Scotland

Introduction

Watson and Moss (1970) summarized the questions we need to ask if we wish to investigate whether social behaviour limits the number of individuals breeding in a population of vertebrates. As ecologists they approached the problem in the most direct fashion emphasizing socially induced mortality, socially induced depression of recruitment, and social restrictions on the exploitation of resources that were themselves not limiting. Their review makes no reference to primates and it is still premature to assess the importance of these factors in the regulation of baboon and macaque populations. Most primatologists have not directly investigated population regulation but have nevertheless raised relevant issues when asking evolutionary questions about the adaptive significance of social behaviour.

Much of the social behaviour shown by primates appears to increase or decrease the Darwinian fitness or inclusive fitness of individuals and may therefore consequently influence the number of individuals breeding in a population. Obviously, other behavioural and environmental factors are also implicated in population regulation but they are largely beyond the scope of this paper. I shall not attempt to assess their relative importance. As an indication of the behaviour I have in mind consider the following examples of how social behaviour may influence the level of predation. A mother who is not vigilant to the approach of predators preying on her young may have a relatively low fitness (compared with other mothers) since she will presumably rear fewer young. If she also shows inadequate care to her sister's offspring she may reduce her inclusive fitness in relation to others in a similar position. When considering the evolution of social behaviour there are three possible evolutionary forces we should keep in mind: individual selection, kin selection, and group selection. The latter will only be considered when the others, involving simpler assumptions more likely to be applicable to wild primates, have been found inadequate (Williams, 1966). I am aware that Wilson (1975, p. 30) calls this the 'fallacy of simplifying the cause'. While I agree with him that one cannot simply reject a process because it seems improbable, the specialized conditions required for group selection (Smith, 1976) make it essential to give the other processes higher priority.

This paper will concentrate on baboon and macaque populations divided into multimale groups. I am most familiar with these and they pose particularly interesting and challenging problems. Instead of discussing the adaptive significance of the full spectrum of behaviour which may be relevant to population regulation, I shall concentrate on critically examining how we can assess whether or not a behaviour is adaptive. In the present context the Darwinian fitness of an individual may be estimated by measuring the individual's relative contribution to the next generation. An estimate of an individual's inclusive fitness may be obtained by measuring the relative contribution of the individual's relatives to the next generation.

Multimale groups of baboons and macaques

Multimale groups are large groups containing a full cross-section of the age–sex classes in the population including several functionally reproductive adult males (Eisenberg *et al.*, 1972). Multimale groups are characteristic of common baboons (e.g. *Papio anubis*, *P. cynocephalus*, *P. ursinus*) (Altmann and Altmann, 1970) and the macaques (Jolly, 1972). In both *Papio* and *Macaca* the whole population is divided into these groups apart from the odd solitary adult male (Altmann and Altmann, 1970; Lindburg, 1971). The proportion of adult males to adult females varies somewhat from species to species and population to population. Deviations from unity are usually in the direction of more females than males (Crook, 1970a). Females remain in their natal group apart from cases of group fission and exceptional transfers between groups. By contrast subadult and adult male baboons frequently transfer between groups (Altmann and Altmann, 1970; Hausfater, 1975; Packer, 1975; Rowell, 1966) and there are similar records for *M. mulatta* (Lindburg, 1969; Boelkins and Wilson, 1972; Missakian, 1973). Long-term data for *P. anubis* reveals that all males change groups at least once (Packer, 1975) and the same is true of Cayo Santiago rhesus (Missakian, 1973).

The genetic relationship between individuals in multimale groups

Can monkeys recognize their genetic relationship with other group members? (By 'recognize' I do not mean conscious awareness of relationships. There is no need to assert, for example, that a monkey thinks 'this is my maternal half-sister'. We need to merely note that animals with a particular genetic relationship have a characteristic behavioural relationship.) It is important to consider this question since a monkey could influence its Darwinian fitness or inclusive fitness by distributing its behaviour in relation to genetic relationships. This is because the degree of relatedness indicates the probability that particular genes (e.g. those controlling altruistic behaviour) are held in common (Partridge and Nunney, in press). If recognition is impossible then selection may favour quite different behavioural strategies. Males might play little part in rearing and concentrate instead on fertilizing as many females as possible. Alternatively, if survival of young is jeopardised by lack of male-care, then we would expect, under some circumstances, to see males staying with the females they have fertilized (Goss-Custard *et al.*, 1972; Clutton-Brock and Harvey, 1976). In a promiscuous society a male may help unrelated young as well as his own and as a result the average gain from a given altruistic act may be lower than in monogamous or polygamous societies. One might therefore expect a male in a multimale group to risk less when helping young than a monogamous or polygamous male.

The importance of matrilineal kinship in determining individual behaviour and relationships has been established by long-term studies of *M. mulatta* and *M. fuscata*. Mothers recognize their young and *vice versa*. Monkeys also recognize their maternal half-siblings and possibly matrilineal relatives as well. Matrilineal kinship influences a wide variety of behaviour including rank, coalitions, grooming partners, ease of transfer between groups, and group division (for references see Crook, 1970a; Hinde, 1974; Jolly, 1972; Clutton-Brock and Harvey, 1976; Wilson, 1975). In spite of the superficial similarity between *Macaca* and *Papio* recent evidence suggests that we must be cautious when generalizing results and speculative theories from one genus to another. For example, males transferring between *M. mulatta* groups tend to transfer into groups already entered by brothers (Boelkins and Wilson, 1972), but in *P. anubis* groups the adult males are not usually closely related (Packer, 1977).

Since the females rarely transfer between groups a multimale group can be visualized as an association of several matrilineal genealogies, each based on a founder female, plus any adult males transferred in from other groups. For example, Loy (1971) found that Cayo Santiago rhesus Group F contained 62 animals. Sixty of the animals were from six matrilineal genealogies, of 5-13 monkeys each, and two adult males had transferred in from other groups. Since females do not move between groups there is a good chance that the founder females are themselves related. For obvious reasons we tend to concentrate on matrilineal relationships but the question of paternity cannot be ignored. The presence of two or more functionally reproductive males in a multimale group results in what I have called the 'paternity problem' (Deag, 1974). There is no evidence that males recognize their offspring and that monkeys recognize their fathers or paternal half-siblings. A similar situation exists in other societies, such as in lions (Bertram, 1976).

Superficially the paternity problem seems real enough but perhaps we should be cautious until we have the opportunity of relating all preferred associations seen in groups to paternal, and not just maternal genetic relationships. What methods might promiscuously breeding males use to help recognize their young? We can discount the possibility of males judging relatedness on the basis of phenotypic similarity to themselves. This is unlikely to provide a reliable index. It also assumes self-recognition, something that macaques (in contrast to chimpanzees) apparently do not have (Gallup, 1970), and the ability of one animal to compare its own and another's appearance

(Partridge and Nunney, in press). Another way for a male to identify his young would be for him to remember which females he copulated with on their most fertile days. We know that males can recognize these days since male–male competition for females is highest at these times. This solution would only be possible where one male has sole access to a female for a day or more. This is sometimes the case in *P. anubis* and *P. cynocephalus* (DeVore, 1965; Collins, personal communication), *M. fuscata* (Tokuda, 1961-62), and *M. mulatta* (Lindburg, 1971) but exceptions are common in each of these species. A further problem is that we do not know whether baboons and macaques can keep a mental list of their pairings. An alternative method, that does not demand this, would be for a male to preferentially copulate with those females with whom he has a special relationship outside the mating context, or vice versa. He would then stand most chance of associating preferentially with his young (Hrdy, 1976). Associations of this type are seen in *P. anubis* and *P. cynocephalus* but there are also clear exceptions (Ransom and Ransom, 1971; Collins, personal communication). I have only considered the question of fathers recognizing their sons. Bertram (1976) looks at other patrilineal relationships and also discusses whether mothers and fathers may have a conflict of interest in the recognition of paternity.

This issue will soon move from speculation to fact when adequate biochemical methods of paternity exclusion become available. In the meantime we should make more use of the available matrilineal information. Average minimal and maximal relationships can be calculated (Clutton-Brock and Harvey, 1976) and if reasonable assumptions can be made about paternity, a more detailed analysis is possible as has been demonstrated by Bertram (1976). Before adequate calculations can be made we do, however, need better information on several points concerning males. What is the relatedness of the males transferring into the same multimale group; what are the effects of group changing, social rank, and the number of competing males on a male's fitness; and what is the relative importance of a male's copulations in his natal group (see Loy, 1971) versus those occurring after transfer?

Before considering some of the implications for individual behaviour of the coexistance of several reproductive males in a group, it is necessary to look critically at the primatologists' approach to investigating the adaptive significance of social behaviour. This is done in the following two sections.

Approaches used to investigate the adaptive significance of primate social behaviour

Does an animal's behaviour increase its Darwinian fitness or inclusive fitness relative to alternative strategies? If the answer to this question is yes, we may call the behaviour adaptive and refer to its function. (Note that I am asking evolutionary questions; proximate mechanisms in development and communication are largely ignored as is the possibility of cultural evolution.)

In studies of primates and other mammals the adaptive significance of social behaviour has been investigated at two levels (Clutton-Brock and Harvey, 1976). Although I have separated these out for clarity it should be noted that they are often used in conjunction with each other.

The first involves examining inter- or intra-specific differences in social organization and relating them to estimated ecological selection pressures (e.g. Crook and Gartlan, 1966; Crook, 1970a) and to estimated social selection pressures such as the ability to act cooperatively (Crook, 1971) and sexual selection (Crook, 1970b; Denham, 1971; Goss-Custard *et al.*, 1972). The arguments developed usually rely on *a posteriori* reasoning and the formation of hypotheses (consistent with current data) about the relationship between different ecological and social variables (Altmann, 1974). Some studies imply that it is meaningful to discuss the adaptive significance of different patterns of social organization (Crook, 1970b; Kummer, 1971). This leaves one open to accusations of group selectionism unless it is emphasized that the patterns seen are a consequence of the adaptive behaviour of individuals (Goss-Custard *et al.*, 1972; Deag, 1974, 1977; Clutton-Brock and Harvey, 1976).

The quality of the factual basis for these arguments varies considerably. Early theories (Crook and Gartlan, 1966; Denham, 1971) relied on quite gross ecological and social organization data which lacked detail of both the current position and of long-term changes (Rowell, 1967). Comparisons are now based on finer information (e.g. Clutton-Brock, 1974) and it is clear that intra-specific comparisons will be particularly valuable (Crook, 1970a; Richard, 1974). At the same time it is important to realize that there are very few top quality demographic studies of monkey populations. These are obviously essential if we are to assess the adaptive significance of social behaviour and to understand the role of social factors in population regulation. An exception is the baboon ecology study by Altmann and Altmann (1970) but even this lacks information on long-term changes. As a consequence of this lack of solid evidence most theoretical treatments depend on inferring the major environmental selection pressures and their relative importance. In seasonal environments, where conditions may fluctuate from year to year or over longer periods, an observer may remain ignorant of the dramatic effects of bad years. For example, in the Amboseli a change in the water table resulted in a salting of the environment. Floral changes and a decrease in the baboon and vervet monkey *Cercopithecus aethiops* populations followed in a way that had not been anticipated when the primates were initially studied (Altmann, 1974; Western and Van Praet, 1973; Struhsaker, 1973). In general, primatologists have avoided declining populations by searching, not unreasonably, for study sites where there are lots of monkeys (Rowell, 1967). Social selection pressures are similarly inferred from a combination of field observations and deducing the effects of individual selection and kin selection. This approach is subjected to further criticisms below.

At the second level, similar methods are applied (intra or inter-specifically) to individual behaviour patterns or the relationships between individuals, and not to whole patterns of social organization or gross categories of behaviour. Recent applications include examination of several types of primate social behaviour, especially agonistic behaviour (Deag, 1974, 1977; Clutton-Brock and Harvey, 1976). In its most flagrant form this approach involves observing a behaviour and then arguing, in the most parsimonious manner available, whether or not it is adaptive. This is decided by judging whether or not it leads to an increase in fitness or inclusive fitness. For the reasons discussed below, this is too simplistic and may be misleading. When applied more critically this approach relies on noting that an animal shows a behaviour and then asking directly 'what selective pressures are acting on it to make it show this behaviour to a greater or lesser extent?' (Bertram, 1976, p. 297). In his study of lions (which, as he points out, raises issues clearly relevant to multimale primate groups) Bertram emphasized kin selection and estimated the degree of relatedness between members of a multimale group. The distribution of social acts (e.g. communal suckling and male tolerance towards cubs at kills) is then related to the animals' degree of relatedness. When an act benefits a close relative (high degree of relatedness) at low cost to the performer (estimated in terms of future reproductive success) it is concluded that kin selection may be involved in its evolution. The interaction between selection pressures is recognized and further selection pressures are invoked as necessary. Packer (1977) used a similar approach in his investigation of reciprocal coalitions between adult male baboons. In a coalition there are two roles, soliciting help and giving aid, and in the most interesting cases the helper was left fighting a male consort while the soliciting male took the latter's female. Packer argued that the males are not closely related and that this excludes any interpretation based on kin selection. The results are therefore interpreted on the basis of individual selection by means of Trivers' (1971) theory of reciprocal altruism.

A complicating factor, the presence of phylogenetic constraints must be mentioned here. There is a tendency to interpret everything we see as an adaptation to current conditions but we know so little about past selection pressures and the speed of evolutionary change, that we should be cautious before jumping to this conclusion. Each animal brings to the present time characteristics which evolved during the past history of its species. These characters provide the basis for future evolutionary change and must to some extent constrain it. For example, the superficial similarity in the social organization of macaques, baboons, and mangabeys probably owes more to their common ancestry than to some common characteristic of their present environment (Chalmers, 1968, 1973; Struhsaker, 1969; Goss-Custard *et al.*, 1972). Such an argument does not, of course, exclude the possibility of there being more subtle interspecific differences which can be related to current selection pressures. Under conditions of rapid environmental change (such as may be due to man) some characters may be maladaptive. In such cases we should see selection against them and we should be prepared to examine their effect in the same way that we examine apparently adaptive traits (see below). Rowell (1972) provides an example: active vigilance has apparently decreased in areas where vigilant monkeys have been targets for hunters with modern firearms.

The phylogenetic constraint issue is also relevant for another reason. When a search for the adaptive significance of social behaviour fails, a convenient retreat is to be found in explaining it as an adaptation to past conditions. For example, having shown that the social organization of *Theropithecus gelada* is inadequately explained by the arguments previously used, Dunbar and Dunbar (1975) revert to arguments based on Plio-Pleistocene conditions. Their hypothesis may be right but it is impossible to test.

The above methods for investigating the adaptive significance of social behaviour involve further problems especially in the types of reasoning used. It is becoming increasingly accepted that one must attempt to interpret the adaptive significance of primate social behaviour on the basis of individual selection before invoking kin selection or group selection (Goss-Custard *et al*, 1972; Deag, 1974, 1977; Clutton-Brock and Harvey, 1976). This point will not be laboured further. A further danger, the selective choice of examples to support a hypothesis, will not be considered. In fact I admit to selectively choosing the following examples to support my case and I apologise to the authors for taking them out of context. It is not my intention to generally disparage their work and I acknowledge using similar methods myself (Deag, 1974, 1977).

Reliance on a posteriori (*inductive*) *reasoning*

Authors commonly rely on arguing from effect to cause. For example, after the discovery of large multimale groups (the effect to be explained) scientists speculated about the reasons (i.e. the causes) for their existence. Goss-Custard *et al*. (1972 p. 11) proposed that 'the presence of several males may enhance group awareness' and Brown (1975) argued that in baboons various factors (high predation pressure, richer habitats, and intergroup competition for food) may promote larger groups. For a behavioural example consider the following. Female Barbary macaques, *M. sylvanus*, give jabber calls during copulation and these sometimes attract other males who break up the courtship before ejaculation and copulate with the females themselves. Calls given before ejaculation therefore have a potentially disruptive effect on a female's copulation and I have tentatively suggested that they may increase a female's fitness by attracting the highest-ranking male who is free and ready to mate (Deag, 1974). Clutton-Brock and Harvey's hypothesis of mother–infant conflict during subsequent copulations (described below) provides another case. Hypothesis of this sort inevitably abound with qualifying terms such as 'could be', 'it is most likely that', and 'may be' but there is no inherent reason why they cannot be correct (Wilson, 1975). Unfortunately when coupled with the assumption that most behaviour is adaptive, this form of reasoning can 'explain' everything (Lewontin, 1977). It is easily forgotten that the 'explanations' produced are really hypotheses which have been developed around the particular set of data they are then used to 'explain'. Authors should not be surprised when their odd words of caution (in a host of speculation) are quickly forgotten.

In many cases alternative hypotheses are possible and it is easy, intentionally or otherwise, to take the most convenient argument. For instance, starting from the observation that under very competitive situations male lions may be more tolerant than females to cubs at kills, Bertram (1976, p. 291) notes 'It is likely that a female generally can recognise her own offspring' but finds it expedient to add that 'it may be difficult in practice for her to discriminate among cubs during competition among hungry animals at a kill'. Males have a higher estimated degree of relatedness (0.31) to the *average* cub than do mothers (0.18). This leaves the way open for explaining male tolerance at kills by kin selection, even though the degree of relatedness between mothers and offspring is of course 0.5 and there are therefore good grounds (as Bertram later acknowledges) for expecting discrimination.

One further problem is that we do not know how much we legitimately demand of the animal's cognitive abilities. Bertram

(1976) finds it necessary to assume that a lioness (A) when deciding whether to suckle a female relation's (B) cubs, needs to remember whether B suckled A's cubs when they were young. This is the modern anthropomorphism. We simply do not know whether lions or monkeys can remember things like this.

Weak assumptions

When considering a hypothesis it is easy to overlook the underlying assumptions and indeed these may be based on weak or insufficient evidence. Selection pressures have typically been estimated and not measured. For example Crook's hypothesis (1970a, pp. 108–109) concerning the division of primate populations under arid conditions into one-male and all-male groups is based on assumptions concerning predation, seasonal food availability, and the climate. When Dunbar and Dunbar (1975) considered one of the species (*Theropithecus gelada*) in more detail they noted that its environment could not really be classified as arid. They used this and other factors, such as the widespread availability of subterranean rhizomes in the dry season, to upset the whole argument. A general relationship may exist between the occurrence of one-male groups and certain environmental variables but it will only be discovered if selection pressures are measured and not guessed. Clutton-Brock and Harvey (1976, p. 201) provide a behavioural example. They offer an explanation of the occurrence of young adult followers in *Papio hamadryas* harems (Kummer, 1968) on the grounds that 'it is to the advantage of the ageing male to share sexual access to his females with a close relative during his declining years, if this enhances his ability to defend the group and diminishes the chance that the group will be occupied by an unrelated animal after his death'. I am, however, unaware of any evidence that followers are relatives (the available information shows that some at least are not sons, Kummer, 1968, p. 56) and that their presence reduces the chance that another male will displace the harem owner from his group.

Estimation and not measurement of the components of fitness

Most arguments involve assumptions about the effect of a behaviour on fitness or inclusive fitness but it should be noted that the components of fitness have been estimated and not measured. The effects of a behaviour on fitness or inclusive fitness are usually deduced on the basis of theories of individual and kin selection and in some cases may be supported by calculating the degree of relatedness between animals. Because of the problem of establishing paternity these calculations can only be estimates. Bertram (1976) has, however, shown that simple modelling can reveal the relative importance of different factors (e.g. number of adult males) on the degree of relatedness. Indirect ways of comparing the reproductive performance of different males include comparing their access to females at their most fertile time (Hall and DeVore, 1965), their relative frequency of mounting bouts (Loy, 1971), and their relative frequency of ejaculations (Drickamer, 1974). Even if one of these measures turns out to be positively correlated with the number of young reared, it must be measured over the animal's lifetime and not, as is currently the case, over a short observation period (Goss-Custard *et al.*, 1972). A start has been made at counting the number of young produced by individual females over a specified time period (Dunbar and Dunbar, 1977). This technique could be used more frequently and extended to matrilineal kin as well. A much better estimate is, however, the lifetime production of young. We will have to wait for biochemical measures of paternity before this information becomes available for males. Estimating inclusive fitness will require measuring the survival and reproduction of relatives and in the case of males, this will involve following them into other groups.

At their worst, hypotheses based on *a posteriori* reasoning are impossible to test. At their best they have weak predictive power and often do not lend themselves to testing since they are formulated around the available information. Speculating about the adaptive significance of social behaviour has become quite a game. Indeed, when played without restraint its (pseudo) explanatory powers make it almost addictive. But, like all games in which the rules can be changed as you go along, it clearly does have problems. The point has come where we must slow down what I believe to be pseudoprogress to get closer to the truth.

Measuring the adaptive significance of social behaviour

One way to make our arguments more rigorous is to emphasize *a priori* (deductive) reasoning in which one argues from cause to effect (Popper, 1968). In principle the starting point is a hypothesis based on evolutionary theory which relates selection pressures to aspects of the social system or the behaviour in question. For instance Clutton-Brock and Harvey (1976, p. 207), considered the theory of kin selection and predicted that (in the context of intra-group agonistic behaviour) 'the amount of tolerance extended to different relatives would correlate with their degree of genetic similarity'. The prediction was then tested with data from *M. fuscata*. In practice, hypctheses are formulated within the context of previous observations and it is difficult to see to what extent inductive or deductive reasoning has been used. For example, consider the following hypothesis which is typical of several formulated by Crook (1970a, p. 125): 'Populations overcrowded in relation to environmental commodities show an increased male exclusion from reproductive groups'. To what extent is the exclusion of males a prediction based on ecological and evolutionary theory or alternatively a summary of what is already known? Hypotheses are typically formulated with reference to one selection pressure but the obvious interaction of factors is usually acknowledged (e.g. Crook and Gartlan, 1966). As more data become available hypotheses are refined to take account of new facts. For instance the discovery of one-male groups in forest monkeys required reassessment (Crook, 1970a) of the hypothesis that one-male reproductive units occur in 'habitats in which food supplies are less abundant, and, at least seasonally, more sparsely and infrequently distributed in the environment' (Crook and Gartlan, 1966, p. 1201).

Wilson (1975) called this combination of logical reasoning and successive modification of hypotheses the 'advocacy method'. Emphasizing *a priori* instead of *a posteriori* reasoning helps, but note that it does not necessarily free one from using arguments of convenience, weak assumptions, or from equating hypotheses with explanations.

The procedure of strong inference

One way forward is to use the procedure of strong inference (Platt, 1964) as suggested by Wilson (1975). The essence of this method is to develop *alternative* hypotheses which are not an end in themselves; their predictions are tested with data that are collected for the purpose. The application of the procedure is illustrated below but it may be helpful if I first summarize the steps to use in the present context.

(*a*) Construct alternative (i.e. competing and not compatable) hypotheses relating selection pressures to inter-individual behavioral variability. These hypotheses cannot be tested directly. For each hypothesis we deduce refutable predictions (Popper, 1968) expressing the relationship between specific parameters that can be observed, measured, and, in appropriate cases, experimentally manipulated. In many cases the alternative to a hypothesis is the direct negative but there are advantages (Platt, 1964) in establishing multiple alternative hypotheses, especially in field biology where long-term observations must be made. The data necessary for testing predictions from alternative hypotheses may be collected simultaneously and a higher rate of scientific progress achieved.

(*b*) Make thorough quantitative and qualitative observations on the behaviour in question and establish adequate means of measuring the performance of recognized individuals at the behaviour. The immediate consequence (or more usually, the multiple interlinked consequences) of the behaviour must be thoroughly understood.

(*c*) Find an adequate sample of individuals who vary for the behaviour in question and who are otherwise comparable. We must move away from thinking about traits in general (e.g. all mothers give maternal care) to examining behavioural variability (Hinde, 1975). (Mothers may, for example, differ in the extent to which they restrain their infants and permit them to interact with other group members.)

(*d*) Design and conduct a set of observations or experiments to test as cleanly as possible the predictions of each hypothesis. In the present context it will be necessary at some stage to compare the fitness of animals who vary for the behaviour in question. The emphasis is on measuring the components of fitness rather than estimating them indirectly.

(*e*) If a prediction is refuted the hypothesis must be rejected. The process must then be restarted by establishing new alternative hypotheses and testing their predictions.

The power and demands of this approach can be illustrated by considering an *a posteriori* hypothesis from Clutton-Brock and Harvey (1976, p. 202). They suggested that 'the frequent attacks made by infants on their mother's subsequent consorts as in orangs (*Pongo pygmaeus*) and chimps (*Pan troglodytes*) (Mackinnon, 1974; Lawick-Goodall, 1968) can be explained' in terms of a conflict of interests. 'By such behaviour, infants could occasionally prevent successful insemination, thus prolonging the period of maternal support, though evidence on this point is lacking.' Applying the method of strong inference our hypothesis is: There are theoretical grounds (which, for the purpose of this example, I will consider to be sound) for believing that parent–offspring conflict may exist (Trivers, 1974). Is the behaviour in question a case of this? For a given population of a given species, and taking care to get an adequate sample at each step, we could proceed to test the following predictions. The initial predictions are concerned with establishing the behavioural phenomenon. This is an essential first step since there would be no point in laboriously testing the later predictions if the early ones are refuted.

Prediction 1. The infant's harassing behaviour is directed at its mother and her consort and apart from isolated cases is not directed at other consorting pairs. (This prediction must be posed because in some species unrelated females may be harassed (Gouzoules, 1974).)

Prediction 2. The infant's behaviour affects the mother's behaviour by prematurely terminating her copulations.

Prediction 3. Mothers that are harassed have increased interbirth intervals and hence conceive fewer young. (It is necessary to make this prediction since a causal link between (2) and (3) is by no means certain.)

If (3) is true there may be grounds for believing that the infants' harassing behaviour lowers the mothers' fitness. This is, however, by no means certain since, for example, offspring born further apart may be more healthy. We must therefore continue and estimate fitness.

Prediction 4. Mothers that are harassed have a lower lifetime reproductive success than those who are not.

Prediction 5. Infants that harass their mothers have a higher lifetime reproductive success than those who do not.

If (4) and (5) are true then we may conclude that harassing during copulations is an expression of the parent–offspring conflict phenomenon and one might therefore reasonably expect to find mothers adopting counter-harassment strategies. Note that one could not conclude from this sequence of hypothesis testing that this behaviour was the only expression of mother–infant conflict or that this was the only selection pressure acting on the infants' harassing behaviour.

I do not underestimate the difficulty of this rigorous approach. Frankly we have to make the choice between admitting the enormity of our aims and accepting slow but intellectually honest progress, or being satisfied with imagined progress based on a convenient set of pseudo-explanations. Although the complexity and diversity of their social behaviour makes primates attractive, their long lives make them quite inappropriate for this type of study. The ease with which the advocacy method arrives at its conclusions has resulted in hypotheses being founded on extremely brief (even anecdotal) observations. We will have to be much surer of our observations before embarking on a long-term project using strong inference and measuring lifetime reproductive success.

Problems caused by the interaction of factors

We are often interested in one particular behaviour but is it ever feasible to measure the effect of a specific behaviour on fitness (Hinde, 1975)? The behaviour of individuals in a multi-male primate group interacts complexly both currently and during social development. In such a complex social environment will it ever be possible to find enough animals who vary for the behaviour in question but who are otherwise comparable? This is unlikely and to surmount the problem it will be necessary to use multivariate statistical techniques. A less satisfactory approach may be to avoid asking questions at such a fine level. Lifetime behaviour strategies could be compared (e.g. high-ranking for life versus low ranking for life) but it would still be necessary to control for other differences between individuals. The construction of models may help us to predict the relative importance of different factors but they will not absolve us from the need to test hypotheses with empirical data.

A second problem is produced by the inevitable interaction of selection pressures acting on a behaviour. Examples will be found in discussions of female-care (i.e. care of infants by females other than their mothers: Deag and Crook, 1971; Wilson, 1975; Hrdy, 1976) and male-care (Deag and Crook, 1971; Mitchell and Brandt, 1972; Deag, 1974; Wilson, 1975; Hrdy, 1976). These reports include numerous hypotheses and amass the evidence for each. Female-care may provide practice in maternal-care for young females, protection of the infant (often a sibling or other close kin), ally the carer with mothers of high rank, as well as benefiting the mothers (Wilson, 1975; Hrdy, 1976). Imagine that in a particular population the relationship between these possible selection pressures and fitness had been tested by the procedure of strong inference. It is highly likely that one would find that several selection pressures were operating. The problem therefore remains, what is the relative importance of the different selection pressures? There is little hope of experimentally manipulating such complex systems to help find the answer. In my opinion there is little point in embarking on the whole process if one cannot solve this problem at the end. This problem of interacting selection pressures is of course exaggerated when the procedure of strong inference has not been used and the arguments involved are speculative and subject to the drawbacks listed earlier. Crook *et al.* (1976) considered that the development of mathematical models will play a major role in the solution of this problem.

Finally we should be prepared to investigate whether apparently adaptive traits are passed to subsequent generations. This will be difficult to study in primates owing to their long generation time, the complex interaction of inheritance and experience during development, and the possibility of cultural inheritance. The success of studies investigating the relationship between the ranks of mothers and their offspring (see p. 86) provides an indication of what might be achieved.

Adult male social behaviour and predation

In view of the difficulties outlined above, is there any point in continuing investigations into the adaptive significance of primate social behaviour? I believe that there is, so long as we are aware of the problems. Measuring the lifetime reproductive success of individuals and their relatives will of course remain difficult. However, I believe that we can make progress if we clearly set out our hypotheses and then proceed to systematically test their predictions. Field experiments must be used where appropriate to control for extrinsic variables or to increase our sample size of what are often rare events (Hall, 1960; Hinde, 1975). This empirical approach will be illustrated by reference to one problem area—the response of adult males to predators. The relative importance of predation in limiting the number of animals breeding in macaque and baboon populations is unknown. Primatologists have, however, speculated extensively about the adaptive significance of presumed anti-predator behaviour and therefore assume that mortality or other interference due to predators is an important selection pressure. Some of the proposed hypotheses will be examined and refined and I shall consider what information is required to test some of their predictions. For brevity a comprehensive sequence of alternative hypotheses, predictions and tests will not be devised.

There is considerable evidence of non-human predation of common baboons in several habitats (Lawick-Goodall, 1968; Altmann and Altmann, 1970; Altmann, 1974; Saayman, 1971; Teleki, 1973; Rhine and Owens, 1972). Altmann and Altmann (1970) give most information on estimated death rates but could not calculate the predation rate since some animals disappeared when observations were not being made. Little information is available about predation of macaques. Domestic dogs are reported attacking *M. mulatta, M. radiata,* and *M. sylvanus* (Lindburg, 1971; Simonds, 1965; Deag, 1974) and actually killing *M. sinica* (Dittus, 1975). Less information is available for wild predators. Leopards, a reputed predator of *M. sylvanus*, are no longer seen in the Atlas but I saw lone jackals *Canis aureus* unsuccessfully hunt the monkeys on five occasions (Deag, 1974). Tigers prey on *M. mulatta* in India (Lindburg, 1971), and in the United States an imported group of *M. fuscata* housed in a large enclosure, has been preyed upon by bobcats *Lynx rufus* (Gouzoules *et al.*, 1975).

Primatologists, on the basis of observations of common baboons in savanna, concluded that the multimale group is in part an anti-predator device (DeVore and Washburn, 1963; Crook and Gartlan, 1966). This argument was extended to macaques in woodland and forest habitats where predation pressure may be much lower (Goss-Custard *et al.*, 1972) and where one would expect less emphasis on confrontation with predators, since rapid escape is possible (Deag, 1974). The whole issue is, however, characterized by its lack of solid data. In fact Eisenberg *et al.* (1972, p. 872) considered that 'no single aspect of primate field studies has less supportive data than the generalizations concerning the survival value of the various presumed anti-predator mechanisms'. It is pointless to extend this debate further. Asked in this general way, 'are multimale groups in part an anti-predator strategy for the individuals that live in them?', the question cannot be answered. A better approach is to consider more specific hypotheses.

Some hypotheses on adult male involvement in anti-predator behaviour

1. Goss-Custard *et al.* (1972) argued that one reason for several males living in the same group is the *individual* advantage they gain from cooperating against predators. Depending on the situation, such cooperation may be true cooperation, in the sense that the males coordinate their behaviour towards an immediate common end, or it may be an indirect result of several individuals working on their own. In a multimale group each male would therefore benefit from the anti-predator behaviour of its fellows. Balanced against this and other advantages, is the disadvantage to some males of sharing access to females (Goss-Custard *et al.*, 1972; Clutton-Brock and Harvey, 1976).
2. Most studies have found that adult males figure prominently in anti-predator behaviour. This has been explained on the grounds that the males are protecting their young. It would, however, be interesting to know how much help is given to unrelated animals as a consequence of the paternity problem.
3. Compared with adult males, other age–sex classes (e.g. lactating females, subadult males, other non-adults) probably participate less frequently in anti-predator behaviour (see below). It is argued that for these animals, the potential costs of some dangerous situations may outweigh the benefits (Deag. 1974).
4. A finely tuned model would predict that adult males should

participate in anti-predator behaviour in proportion to their relative mating success so far. High-ranking adult males might therefore be relatively more involved in protection (Hall, 1960; Hrdy, 1976). Such a prediction would need modification if breeding males were relatives or if a male had female relatives breeding in his own group.

The pre-eminent participation of adult males may be facilitated by their more powerful build but the extent to which the latter is attributable to sexual selection or success against predators is debatable (Crook, 1972). Hall (1964) considered that when highly aroused adult male baboons have a strong social facilitive tendency to act together against predators. Goss-Custard *et al.* (1972, p. 11) noted that 'males may be more alert to potential danger since they are not involved in direct caring for the young'. Proximate mechanisms like these will not be discussed further, except to say that in the latter case other aspects of the animals' time–energy budget (e.g. food requirements) would also have to be considered.

Unfortunately this is still an inadequate level at which to speculate about anti-predator strategies. Predictions must be made from these hypotheses about individual differences in the performance of specific types of anti-predator behaviour. The anti-predator behaviour of adult males will therefore be summarized and the outstanding questions then noted.

The 'anti-predator' behaviour of adult males

1. *Adult male 'lookouts'.* Adult male *P. anubis* and *M. sylvanus* sit on vantage points looking around (Rowell, 1972; Deag, 1974). This behaviour is distinguished from active vigilance (see below). Although I have no quantitative data my impression was that 'lookout' behaviour is largely restricted to adult and subadult males. By banging or whistling I was able to show that the animals were alert and would turn to face the sound source (Deag, 1974). Rowell (1972, p. 43) reported that if one of these males 'saw anything unusual, he would investigate or avoid, and his action would be immediately observed by other watching males, who would come to reinforce his investigation, or shepherd the troop away as appropriate'. 'Lookout' behaviour may therefore give way to active vigilance. Hall (1960) did not see 'lookout' behaviour in *P. ursinus* but considered that this

was due to the lack of predation in his study area. On the other hand, in *P. cynocephalus* the behaviour may be rare even though predators are present (Collins, personal communication).

In *P. anubis* and *M. sylvanus* a similar behaviour may be seen when males sit looking towards another group (Collins, personal communication; Deag, 1974) and such instances must be excluded from those considered here.

2. *Adult males in active vigilance.* Even the quietest approach to an unhabituated Barbary macaque group is soon detected and barks given. If the observer maintains position, and does not rapidly approach the group, an adult male normally moves to a vantage point, sits and watches him. It may bark. Vigilance can be continued for many minutes, the rest of the group typically moving out of sight behind the vigilant animal (Deag, 1974). Although details vary, a similar behaviour is reported in *M. mulatta* (Lindburg, 1971), *M. speciosa* (Bertrand, 1969), *P. ursinus* (Hall, 1960; Stoltz and Saayman, 1970) and *P. anubis* (Rowell, 1966, 1972).

In one group of *M. sylvanus* the three adult males were responsible for 74% of the instances of active vigilance even though they constituted only 12% of the group. The remaining instances were by subadult males and adult, subadult, and juvenile females (Deag, 1974). The other studies mentioned above reported similar findings. In *M. sylvanus* there seemed to be no true cooperation between the vigilant males. In contrast, Rowell (1972) reported antiphonal calling between the vigilant male and a male moving in the progession away from the danger.

3. *Adult males confronting predators.* Only cases where animals confront predators (i.e. stand their ground, lunge at, chase, or attack) are considered here. So-called protective formations and the defence of young are discussed below. Confrontations vary in intensity and may involve animals other than adult males. Key observations from the modern baboon literature include the following. (*a*) Three cheetahs leaving after being threatened by an adult male (DeVore and Washburn, 1963). (*b*) An adult male chasing off jackals; an adult male attacking, and juveniles threatening a tawny eagle; and leopards being threatened,

chased, and attacked by a mass of individuals including adult males and adult females with young clinging to their bellies (Altmann and Altmann, 1970). (*c*) Lions being mobbed by prolonged calling by more or less a whole group (Saayman, 1971). (*d*) Adult male baboons have been seen to kill domestic dogs that chased a group (Stoltz and Saayman, 1970). Baboons of all age–sex classes will on occasion flee from predators such as lions and leopards (see below).

Lindburg (1971) found that adult male *M. mulatta* were usually responsible for threatening or chasing away domestic dogs. Females also participated but tended to be involved in more minor incidents. Japanese macaques of all age–sex classes threaten hunting bobcats but adult males tend to be more active in driving them off (Gouzoules *et al.*, 1975). Barbary macaques do not confront hunting jackals or dogs—all age–sex classes escape by climbing trees. Adult males were usually the first to descend after the dogs left, and on some occasions they moved out towards where the dogs had gone and stood looking before settling down to feed (Deag, 1974). The early descent of adult males could be explained at a proximal level by assuming them to be the least fearful animals (see below). This would not explain their movement towards where the predator disappeared and their vigilance behaviour.

On three occasions several Barbary macaques (adult, subadult, and juvenile males, and adult females) threatened me. One incident, triggered by working too close to the animals, was initiated by two adult males but other age–sex classes soon joined in (Deag, 1974). Similar incidents, in which adult males played a major but not exclusive part, are reported for *M. mulatta* (Koford, 1963), *M. fuscata* (Carpenter, 1974), *P. anubis* and *P. cynocephalus* (Collins, personal communication) and *P. ursinus* (Stoltz and Saayman, 1970).

4. *Adult males protecting young from predators*. The extent to which males care or otherwise interact with young has been the subject of considerable interest (Deag and Crook, 1971; Deag, 1974; Hrdy, 1976). In the present context attention will be restricted to the direct protection of young from predators; generalized group defence was mentioned earlier. Protection is most obvious when an observer inadvertently isolates a young

macaque (*M. sylvanus, M. mulatta*) or baboon (*P. ursinus*) in a tree. In *M. sylvanus* the isolated animal screamed loudly and a rescue, involving an approach to the observer, was accomplished by the mother or an adult male (Deag, 1974). Under similar circumstances Lindburg (1971) and Stoltz and Saayman (1970) respectively reported the involvement of adult males and the whole group. When infant Barbary macaques screamed at other times they were approached by members of all age–sex classes as well as their mother (Deag, 1974). When a bobcat seized a young Japanese macaque, all age–sex classes mobbed the predator but the two top-ranking adult males (who were consorting at the time) ignored the incident (Gouzoules *et al.*, 1975). When a person makes a close approach to the same species or attempts to catch infants the adult males advance threateningly (Eaton, 1976) or hold infants protectively while threatening the approacher (Itani, 1959). In some species, male-care relationships (Deag and Crook, 1971) are established between specific male–infant pairs (Ransom and Ransom, 1971; Deag, 1974; Hrdy, 1976). There is, however, insufficient evidence to show whether males preferentially rescue or protect those infants with whom they have a special relationship.

5. *Adult males 'mobilizing' the group.* Buskirk *et al.* (1974) described how in some groups adult male *P. ursinus* consolidated the group and influenced its movement by herding, chasing, and calling. As a result straying animals rejoined the main part of the group and may therefore have been less susceptible to predation. The authors concluded that the male's behaviour 'probably prevents loss of females available for mating and insures safety of the offspring'. This conclusion is tentative, however, since the control and consequences of the behaviour are poorly understood.

6. *Adult males in 'protective' formations.* Considerable attention has been devoted to the relative spatial positions of individuals of different age–sex classes in moving baboon groups. DeVore and Washburn (1963) and Hall and DeVore (1965) found that *P. anubis* and *P. ursinus* groups moving out into open plains had a clear and invariable progression order. From the front to the back of the group the order was: less dominant adult males and

older juvenile males; pregnant and oestrus adult females and juveniles; dominant adult males; females with infants and young juveniles; adults and other juveniles; adult males. This formation was considered to insure 'maximum protection for the infants and juveniles in the centre of the troop. An approaching predator would first encounter the adult males on the troop's periphery, and then the adult males in the centre, before it could reach defenceless troop members in the centre' (DeVore and Washburn, 1963). Subsequent discussion of these observations has centred on three issues.

(*a*) For a given type of progression (e.g. fast or slow, single file or widely dispersed) in a given context (e.g. on the open plains, in thick bush, approaching sleeping sites or water holes, moving after an alarm) is the spatial distribution of individuals or age–sex classes non-random? If non-randomness is found does the pattern conform to that described by DeVore and Washburn? This problem has been subjected to considerable quantitative analysis. Detailed reviews are available in Rhine (1975) and Collins (in preparation) and will therefore be considered only briefly. Non-quantitative results from three sites (reviewed by Altmann and Altmann, 1970) indicated no special positioning of adult males. There is, however, growing evidence from quantitative studies that, in spite of the constant rearrangement of individuals, adult males (as a class) tend to occur more frequently than expected at the front, back, and sides of progressions (Rhine, 1975; Collins, in preparation). Both Rhine and Collins found marked individual differences in positions: the high-ranking adult males were most frequently at the front, while the old, low-ranking adult males occurred at the back. The central positioning of high-ranking adult males (DeVore and Washburn, 1963) has not been confirmed.

(*b*) Is the non-random positioning of individual adult males the result of individuals adopting particular anti-predator strategies? This issue has been examined critically by Rhine (1975) and Collins (in preparation) but the key questions remain unanswered. Describing the relative spatial positions of individuals will not alone solve this aspect of the problem. DeVore and Washburn's interpretation of their baboon formation is a clear example of an *a posteriori* hypothesis, and it is of course possible to think up other ways in which adult males could position themselves to protect the young, according to the context and their past experience (Rhine, 1975).

(c) Is the non-random positioning of individuals an indirect consequence of other mechanisms? DeVore and Washburn (1963) observed that when a group fled from predators the adult males lagged behind to form a protective screen. Similar formations in response to mildly alarming situations have been recorded by Rowell (1966) and Stoltz and Saayman (1970). Rowell, however, noticed that when the flight stimulus was stronger the adult males fled first leaving the females and young behind. On this interpretation any protective screen formed by the adult males is simply a consequence of them being less fearful. Rhine (1975) extended this idea to account for the position of males in non-flight situations and pointed out that differential fear may be the proximate mechanism for implimenting the evolutionary response to predation. When males flee before females, the concern for individual safety presumably overrides that for the survival of offspring and other kin.

The situation is further complicated by the fact that much of the variation in position of different adult males can be explained by factors quite unrelated to predation pressure. Collins (in preparation) demonstrated how the spatially peripheral position of certain high-ranking males recently transferred into a group reflected, and might be explained by the fact, that they were 'poorly integrated' with other group members. A male's interest in infants or sexual partners also determines his spatial position (Rhine, 1975). It is essential to maintain an open mind on this whole issue until the relative importance of the different selection pressures has been determined. Examination of the spatial distribution of individuals in groups of some macaque species reveals similarities to the pattern described for baboons (Deag, 1974). And yet the importance of predation pressure in this regard is all but ignored; interpretations have largely been concerned with the behavioural relationships between individuals (Deag, 1974). This illustrates yet again how inadequate it is to just assume that you know the adaptive significance of a behaviour without actually setting out to critically investigate the whole problem.

Outstanding questions on adult male social behaviour and predation

Hall (1960) concluded his detailed study of active vigilance by pointing out the inadequacies in our knowledge, a situation that is largely unchanged today. Much of the behaviour discussed above appears adaptive but in no case do we have sufficient evidence to answer the crucial question—what is the relationship between individual differences in behaviour and fitness? Our knowledge of the immediate consequences of the behaviour and of the predators' hunting strategies is also inadequate. Some of the questions we need to ask are now summarized.

(*a*) What are the immediate consequences of the behaviour? For example, are 'lookouts' more successful than other animals at noticing potential predators? Are vigilant animals better than others at warning of further danger or is there a rapid decrement in vigilance performance (Dimond and Lazarus, 1974)? These points could be tested experimentally with dogs as predators.

(*b*) What is the relative participation of the different age–sex classes? Has the participation of adult males been exaggerated because of their large size and eye-catching properties (Collins, personal communication)? How often is the behaviour shown? It is easy to over-emphasize the importance of occasional observations.

(*c*) What risks do the different types of anti-predator behaviour involve? For example, active vigilance may be very risky when the observer is also a hunter (Rowell, 1972). Information is needed on the relative risks of different predators and of the different types of confrontations (e.g. lone threatening versus group mobbing) and how this relates to the participation of different age–sex classes and individuals. What risks are associated with various spatial positions in progressions? How does this change with the context of the progression?

(*d*) What is the relationship between individual differences in the performance of adult males and the number of young (and other relatives) they have living in the group? For example, are there typical strategies for adult males in their natal group, newly transferred adult males, and well-established adult males in non-natal groups?

(*e*) What is the relationship between individual differences in behaviour (especially among adult males) and their fitness and inclusive fitness? Several reports draw attention to the behaviour of the highest-ranking male both in the wild (Altmann and Altmann, 1970; Koford, 1963) and captivity (Bernstein and Sharpe, 1966; Bernstein, 1966; Eaton, 1976). There are, however, exceptions (Gouzoules *et al.*, 1975) and the whole topic requires detailed investigation in the wild, as does the relationship between rank and lifetime reproductive success (Deag, 1977). Hrdy (1976, p. 113) concluded that the 'males most likely to be fathers are apparently those males that also protect and care for infants'. This generalization is premature.

(*f*) How important is cooperation between males and how does this relate to their degree of relatedness and to other instances of cooperation between them?

(g) What is the relationship between the number of adult males in a group and their anti-predator behaviour, and the reproductive success of other group members?

General conclusions and summary

This paper draws attention to the superficial way in which primatologists have generally discussed the adaptive significance of social behaviour. After summarizing the types of arguments used and their inherent problems, a much more critical approach based on established facts is demanded. This is illustrated by examining the so-called anti-predator strategies of adult male baboons and macaques, where there are particularly interesting problems owing to the inability of males to recognize their offspring and vice versa. We must take care to establish the form of the behaviour, the relative participation of individuals from different age–sex classes, and the immediate consequences of the behaviour. Alternative *a priori* hypotheses must be formulated about the adaptive significance of individual differences in behaviour. More attention must be devoted to hypothesis testing, measuring selection pressures and measuring the lifetime reproductive success of individuals. The interaction of different selection pressures and the interaction of different proximate causal factors are discussed. They both produce particularly serious problems for assessing the evolutionary importance of individual differences in behaviour.

Acknowledgments

My field research in Morocco was undertaken during 1968–69 when I was based at the Department of Psychology, University of Bristol. It was financed by Leverhulme Research Awards, the Wenner-Gren Foundation, the University of Bristol, and the Science Research Council. The assistance of l'Administration des Eaux et Forets et de la Conservation des Sols, Rabat, is gratefully acknowledged. I thank the following colleagues for valuable criticism of this paper during its preparation: Mr D. A. Collins, Mr M. Dow, Professor A. W. G. Manning, and Dr L. Partridge. This does not imply that they necessarily agree, with all the views expressed.

References

Altmann, S. A. (1974) Baboons, space, time and energy. *American Zoologist*, **14**, 221–248.

Altmann, S. A. and Altmann, J. (1970) *Baboon Ecology*. Chicago: University of Chicago Press.

Bernstein, I. S. (1966) An investigation into the organization of Pigtail Monkey groups through the use of challenges. *Primates*, **7**, 471–480.

Bernstein, I. S. and Sharpe, L. G. (1966) Social roles in a rhesus monkey group. *Behaviour*, **26**, 91–104.

Bertram, B. C. R. (1976) Kin selection in lions and in evolution. In *Growing Points in Ethology*, ed. Bateson, P. P. G. and Hinde, R. A. pp. 281–301. Cambridge: Cambridge University Press.

Bertrand, M. (1969) The behavioural repertoire of the Stumptail Macaque. *Bibliotheca Primatologica*, **11**, 1–273.

Boelkins, R. C. and Wilson, A. P. (1972) Intergroup social dynamics of the Cayo Santiago rhesus (*Macaca mulatta*) with special reference to changes in group membership by males. *Primates*, **13**, 125–140.

Brown, J. L. (1975) *The Evolution of Behaviour*. New York: Norton.

Buskirk, W. H., Buskirk, R. E., and Hamilton, W. J. (1974) Troop-mobilizing behaviour of adult male chacma baboons. *Folia Primatologica*, **22**, 9–18.

Carpenter, C. R. (1974) Aggressive behavioural systems. In *Primate Aggression, Territoriality and Xenophobia*, ed. Holloway, R. L. pp. 459–496. New York: Academic Press.

Chalmers, N. R. (1968) The social behaviour of free living mangabeys in Uganda. *Folia Primatologica*, **8**, 263–281.

Chalmers, N. R. (1973) Differences in behaviour between some arboreal and terrestrial species of African monkeys. In *Comparative Ecology and Behaviour of Primates*, ed. Michael, R. P. and Crook, J. H. pp. 68–100. London: Academic Press.

Clutton-Brock, T. H. (1974) Primate social organization and ecology. *Nature, London*, **250**, 539–542.

Clutton-Brock, T. H. and Harvey, P. H. (1976) Evolutionary rules and primate societies. In *Growing Points in Ethology*, ed. Bateson, P. P. G. and Hinde, R. A. pp. 195–237. Cambridge: Cambridge University Press.

Collins, D. A. (in preparation) Sexual consort behaviour in yellow and olive baboons in Tanzania. Ph.D. thesis, University of Edinburgh.

Crook. J. H. (1970a) The socio-ecology of primates. In *Social Behaviour in Birds and Mammals*, ed. Crook, J. H. pp. 103–166. London: Academic Press.

Crook, J. H. (1970b) Social organization and the environment: aspects of contemporary social ethology. *Animal Behaviour*, **18**, 197–209.

Crook, J. H. (1971) Sources of cooperation in animals and man. In *Man and*

Beast: Comparative Social Behaviour, ed. Eisenberg, J. F. and Dillon, W. S. pp. 235–260. Washington: Smithsonian Institution Press.

Crook, J. H. (1972) Sexual selection, dimorphism, and social organization in the primates. In *Sexual Selection and the Descent of Man, 1871–1971*, ed. Campbell, B. pp. 231–281. Chicago: Aldine.

Crook, J. H., Ellis, J. E., and Goss-Custard, J. D. (1976) Mammalian social systems: structure and function. *Animal Behaviour*, **24**, 261–274.

Crook, J. H. and Gartlan, J. S. (1966) Evolution of primate societies. *Nature, London*, **210**, 1200–1203.

Deag, J. M. (1974) A study of the social behaviour and ecology of the wild Barbary macaque *Macaca sylvanus*, L. PhD thesis, University of Bristol.

Deag, J. M. (1977) Aggression and submission in monkey societies. *Animal Behaviour*, **25**, **465–474.**

Deag, J. M. and Crook, J. H. (1971) Social behaviour and 'agonistic buffering' in the wild Barbary macaque, *Macaca sylvanus* L. *Folia Primatologica*, **15**, 183–200.

Denham, W. W. (1971) Energy relations and some basic properties of primate social organization. *American Anthropologist*, **73**, 77–95.

DeVore, I. (1965) Male dominance and mating behaviour in baboons. In *Sex and Behaviour*, ed. Beach, F. A. pp. 266–289. New York: Wiley.

DeVore, I. and Washburn, S. L. (1963) Baboon ecology and human evolution, In *African Ecology and Human Evolution*, ed. Howell, F. C. and Bourlière, F. pp. 335–367. Chicago: Viking Fund Publications in Anthropology.

Dimond, S. and Lazarus, J. (1974) The problem of vigilance in animal life. *Brain, Behaviour and Evolution*, **9**, 60–79.

Dittus, W. P. J. (1975) Population dynamics of the Toque Monkey *Macaca sinica*. In *Socioecology and Psychology of Primates*, ed. Tuttle, R. H. pp. 125–151. The Hague: Mouton Publishers.

Drickamer, L. C. (1974) Social rank, observability, and sexual behaviour of rhesus monkeys (*Macaca mulatta*). *Journal of Reproduction and Fertility*, **37**, 117–120.

Dunbar, R. and Dunbar, P. (1975) Social dynamics of Gelada baboons. *Contributions to Primatology*, **6**, 1–157.

Dunbar, R. I. M. and Dunbar, E. P. (1977) Dominance and reproductive success among female gelada baboons. *Nature, London*, **266**, 351–352.

Eaton, G. G. (1976) The social order of Japanese macaques. *Scientific American*, **235**(4), 96–106.

Eisenberg, J. F., Muckenhirn, N. A., and Rudran, R. (1972) The relation between ecology and social structure in primates. *Science*, **176**, 863–874.

Gallup, P. G. (1970) Chimpanzees: self recognition. *Science*, **167**, 86–87.

Gouzoules, H. (1974) Harassment of sexual behaviour in the stumptail macaque, *Macaca arctoides*. *Folia Primatologica*, **22**, 208–217.

Gouzoules, H., Fedigan, L. M., and Fedigan, L. (1975) Response of a transplanted troop of Japanese macaques (*Macaca fuscata*) to Bobcat (*Lynx rufus*) predation. *Primates*, **16**, 335–349.

Goss-Custard, J. D., Dunbar, R. I. M., and Aldrich-Blake, F. P. G. (1972) Survival, mating and rearing strategies in the evolution of primate social structure. *Folia Primatologica*, **17**, 1–19.

Hall, K. R. L. (1960) Social vigilance behaviour in the Chacma baboon, *Papio ursinus*. *Behaviour*, **16**, 261–294.

Hall, K. R. L. (1964) Aggression in monkey and ape societies. In *The Natural History of Aggression*, ed. Carthy, J. D. and Ebling, J. F. pp. 51–64. London: Academic Press.

Hall, K. R. L. and DeVore, I. (1965) Baboon social behaviour. In *Primate Behaviour*, ed. DeVore, I. pp. 53–110. New York: Holt, Rinehart and Winston.

Hausfater, G. (1975) Dominance and reproduction in baboons (*Papio cynocephalus*). A quantitative analysis. *Contributions to Primatology*, **7**, 1–150.

Hinde, R. A. (1974) *Biological Bases of Human Social Behaviour*. New York: McGraw-Hill.

Hinde, R. A. (1975) The concept of function. In *Function and Evolution in Behaviour*, ed. Baerends, G., Beer, C., and Manning, A. pp. 3–15. Oxford: Clarendon Press.

Hrdy, S. B. (1976) Care and exploitation of nonhuman primate infants by conspecifics other than the mother. *Advances in the study of Behaviour*, **6**, 101–158.

Itani, J. (1959) Paternal care in the wild Japanese monkey *Macaca fuscata fuscata*. *Primates*, **2**, 61–93.

Jolly, A. (1972) *The Evolution of Primate Behaviour*. Macmillan: New York.

Koford, C. B. (1963) Group relations in an island colony of rhesus monkeys. In *Primate Social Behaviour*, ed. Southwick, C. H. pp. 136–152. Princeton: Van Nostrand.

Kummer, H. (1968) Social organization of Hamadryas baboons. *Bibliotheca Primatologica*, **6**, 1–189.

Kummer, H. (1971) *Primate Societies*. Chicago: Aldine.

Lawick-Goodall, J. van (1968) The behaviour of free-living chimpanzees in the Gombe Stream Reserve. *Animal Behaviour Monographs*, **1**, 161—311.

Lewontin, R. C. (1977) Caricature of Darwinism. *Nature, London*, **266**, 283–284.

Lindburg, D. G. (1969) Rhesus monkeys: mating season mobility of adult males. *Science*, **166**, 1176–1178.

Lindburg, D. G. (1971) The rhesus monkey in North India: an ecological and behavioural study. In *Primate Behaviour Vol. 2*, ed. Rosenblum, L. A. pp. 1–106. New York: Academic Press.

Loy, J. (1971) Estrous behaviour of free-ranging rhesus monkeys (*Macaca mulatta*). *Primates*, **12**, 1–31.

Mackinnon, J. (1974) The ecology and behaviour of wild orang-utans (*Pongo pygmaeus*). *Animal Behaviour*, **22**, 3–74.

Missakian, E. A. (1973) Genealogical mating activity in free-ranging groups of rhesus monkeys (*Macaca mulatta*) on Cayo Santiago. *Behaviour*, **45**, 225–241.

Mitchell, G. and Brandt, E. M. (1972) Paternal behaviour in primates. In *Primate Socialization*, ed. Poirier, F. E. pp. 173–206. New York: Random House.

Packer, C. (1975) Male transfer in olive baboons. *Nature, London*, **255**, 219–220.

Packer, C. (1977) Reciprocal altruism in *Papio anubis*. *Nature, London*, **265**, 441–443.

Partridge, L. and Nunney, L. (in press) Three generation family conflict: matters arising. *Animal Behaviour*.

Platt, J. R. (1964) Strong inference. *Science*, **146**, 347–353.

Popper, K. R. (1968) *The Logic of Scientific Discovery* (revised edition). London: Hutchinson.

Ransom, T. W. and Ransom, B. S. (1971) Adult male–infant relations among baboons (*Papio anubis*). *Folia Primatologica*, **16**, 179–195.

Rhine, R. J. (1975) The order of movement of yellow baboons (*Papio cynocephalus*). *Folia Primatologica*, **23**, 72–104.

Rhine, R. J. and Owens, N. W. (1972) The order of movement of adult male and black infant baboons (*Papio anubis*) entering and leaving a potentially dangerous clearing. *Folia Primatologica*, **18**, 276–283.

Richard, A. (1974) Intraspecific variation in the social organization and ecology of *Propithecus verreauxi*. *Folia Primatologica*, **22**, 178–207.

Rowell, T. E. (1966) Forest living baboons in Uganda. *Journal of Zoology, London*, **149**, 344–364.

Rowell, T. E. (1967) Variability in the social organization of primates. In *Primate Ethology*, ed. Morris, D. pp. 219–235. London: Weidenfeld and Nicholson.

Rowell, T. E. (1972) *Social Behaviour of Monkeys*. Harmondsworth, Middlesex: Penguin.

Saayman, G. S. (1971) Baboons' responses to predators. *African Wildlife*, **25**, 46–49.

Simonds, P. E. (1965) The Bonnet Macaque in South India. In *Primate Behaviour*, ed. DeVore, I. pp. 175–196. New York, Rinehart and Winston.

Smith, J. M. (1976) Group selection. *Quarterly Review of Biology*, **51**, 277–283.

Stoltz, L. P. and Saayman, G. S. (1970) Ecology and behaviour of baboons in the Northern Transvaal. *Annals Transvaal Museum*, **26**, 99–143.

Struhsaker, T. T. (1969) Correlates of ecology and social organization among African cercopithecines. *Folia Primatologica*, **11**, 80–118.

Struhsaker, T. T. (1973) A recensus of vervet monkeys in the Masai-Ambroseli Game Reserve, Kenya. *Ecology*, **54**, 930–932.

Tokuda, K. (1961–1962) A study of the sexual behaviour in the Japanese monkey troop. *Primates*, **3**, 1–40.

Trivers, R. L. (1971) The evolution of reciprocal altruism. *Quarterly Review of Biology*, **46**, 35–57.

Trivers, R. L. (1974) Parent-offspring conflict. *American Zoologist*, **14**, 249–264.

Teleki, G. (1973) *The Predatory Behaviour of Wild Chimpanzees*. Cranbury, New Jersey: Associated Universities Press.

Watson, A. and Moss, R. (1970) Dominance, spacing behaviour and aggression in relation to population limitation in vertebrates. In *Animal Populations in Relation to their Food Supply*. ed. Watson, A. pp. 167–220. Oxford: British Ecological Society Symposium Number Ten.

Western, D. and Van Praet, C. (1973) Cyclical changes in the habitat and climate of an East African ecosystem. *Nature, London*, **241**, 104–106.

Williams, G. C. (1966) *Adaptation and Natural Selection. A Critique of Current Evolutionary Thought*. Princeton: Princeton University Press.

Wilson, E. O. (1975) *Sociobiology: the New Synthesis*. Cambridge, Massachusetts: Belknap Press.

The place of odours in population processes

D. MICHAEL STODDART

Department of Zoology, University of London King's College, London, England

Introduction

It is only within recent years that much interest has been attached to the behavioural and ecological role of odorous secretions. Partly this has been because man is a microsmatic species which is more attracted to visual patterns and sounds than to ill-defined, and often ill-smelling, odours, and partly because of the difficulty inherent in analysing complex mixtures. Much of the current interest has arisen from the post-war development of a more holistic approach towards animal biology and from the readily available analytical techniques. Many basic anatomical studies on scent-producing apparatus were performed over half a century ago and only now, in the light of behavioural and ecological studies, is their value being revised.

In the terrestrial arthropods, odours appear to have relatively simple chemical compositions and usually have only a single function. Often this is related to mate finding. Among the social hymenoptera and the termites, odours serve to govern the rates and directions of development of colony members. Amphibians have a poorly developed sense of smell. Fish are able to navigate over incredible distances by following chemical cues dissolved in the water. Some reptiles have a well developed olfactory sense but little is known of the biological role of their odours. Few birds have anything more than a rudimentary sense of smell. It is in the mammals that one finds the greatest array of olfactory adornments and development and it is within this class that most attempts have been made to identify and describe the importance and potency of odours (Stoddart, 1976). For this reason this article will deal only with mammalian studies.

This symposium is concerned with the inter-relationships between behaviour and population regulation mechanisms. Although there is only a small amount of evidence yet available which illustrates the role of odour in population processes and in individual behaviour, more is becoming available every day. It seems right, therefore, to examine what is known at this stage and to point out those areas from which more data are urgently needed.

Mention will be made of the studies which illustrate the involvement of odour in reproduction only to point out a difficulty of some proportion. The studies which demonstrated prolonged anoestrus in crowded female mice (van der Lee and Boot, 1956), oestrous cycle synchrony induced by male odour (Whitten, 1959), litter resorption induced by odour of strange male (Bruce, 1960), and family cohesion effected by a maternal odour (Leon and Moltz, 1972) have all been performed in the laboratory under artificial and controlled conditions. Consideration of the social organization of wild mice precludes the development of the conditions continued under artificial conditions. Great caution is needed in the interpretation of these studies. The same *caveat* applies to all studies conducted under laboratory conditions, but to date there have been few others.

Apart from courtship and reproduction, there are three main areas in population ecology in which odours can be seen, or at least presumed, to play an influencing role. These are in the establishment and labelling of social dominance, the demarcation and maintenance of territories and other individual or group living space, and in the transmission of warning or danger signals. All of these aspects of population biology influence future population levels and the ways in which they operate are themselves the result of the impact of the environment and its resource levels upon the population.

It is a point of some importance that odour production is generally more developed in male mammals than in females or juveniles (Schaffer, 1940). Although there are some good examples of male-produced odour being used as a sex attractant (for example the well known odour of male pigs which induces tranquility and mating posture in sows, and the odour of tom-cats which induces copulatory crouching in she-cats) there is little doubt that male mammals usually seek out the females and take the lead in courtship. Unless the male's odour induces the female to produce her attractive odour, and there is no evidence for this, their highly developed odour glands play no primary part in mate attraction, though they may play a secondary part once courtship has started. Furthermore, scent glands occur in many species which have a stable social life with the sexes remaining paired for life. In these species, e.g. Maxwell's duiker *Cephalophus maxwelli* (Ralls, 1969), the male has no need to seek out the female. In other species both sexes bear glands which are active not only during the breeding season but throughout the year; for example, *Oryctolagus cuniculus* (Mykytowycz, 1965), *Petaurus breviceps* (Schultze-Westrum, 1965), *Arvicola terrestris* (Stoddart, 1972), *inter alia*. It must be emphasized that a courtship or bond-maintaining role is not precluded for these odours, but if such roles occur they would seem to be of secondary importance. Studies on behavioural effect of odours have shown that male-produced odours have more effect on other males than on either females or juveniles (e.g. Aleksiuk, 1968; Epple, 1974; von Holst and Buergel-Goodwin, 1975; Jones and Nowell, 1973; Mykytowycz and Dudzinski, 1966; Schultze-Westrum, 1965; Thiessen, 1968).

It is not the object of this paper to discuss scent-marking behaviour. Good reviews will be found in Ralls (1971) and Johnson (1973). Suffice it to say that the frequency and intensity of scent-setting behaviour is generally closely correlated with the stage of development of the scent-producing organ (von Holst and Buergel-Goodwin, 1975; Mykytowycz, 1965; Schultze-Westrum, 1965; Thiessen, 1968). Developmental studies on scent glands and organs have indicated that cycles of glandular activity in males parallel the cycle of testicular activity and are controlled by levels of blood steroids (von Holst and Buergel-Goodwin, 1975; Martan, 1962; Stoddart, 1972; Thiessen, 1968; Vrtis, 1930). Not only is scent production increased during the breeding season but the chemical composition appears to change (Hesterman and Mykytowycz, 1968; Stoddart *et al.*, 1975). These events, which are related to the period of offspring production and enhanced social interaction, highlight the social nature of odour production (Stoddart, 1977).

The establishment and labelling of social dominance

Many workers have observed an increase in the frequency of scent-marking behaviour associated with a rise in social dominance status of the marker or when confronted by an aggressive neighbour (Beauchamp, 1974; Drikamer *et al.*, 1973; Epple, 1974; Hesterman and Mykytowycz, 1968; von Holst and Buergel-Goodwin, 1975; Jones and Nowell, 1973; Mykytowycz, 1965; Mykytowycz and Dudzinski, 1966; Ralls, 1969; Schultze-Westrum, 1965; Thiessen, 1968), but few have attempted to examine the relationship between odour quality and dominance.

Although numerous anecdotal observations on seasonal and sexual differences in mammalian body odour exist it was Hesterman and Mykytowycz in 1968 who first showed experimentally that such differences occur in the rabbit. They employed a panel of 'sniffers' who were asked to say which of two vials contained the more intense odour. Care was taken to ensure the concentrations were kept standard. The result of this investigation was that human judges found the odour of marking pellets, which are coated in anal gland secretion, to be stronger in dominant males than in subordinate males. Odour intensity was always higher in males than in females and increased with age. Additionally, odour was more intense during the breeding season. Mykytowycz and Dudzinski (1966) showed that the frequency of chinning behaviour increased with rise in dominance status and the size of the submandibular, anal, and inguinal glands also increased. So not only do dominant rabbits, of both sexes, produce more odorous secretions but the quality of these secretions also changes. Goodrich and Mykytowycz (1972) showed that the composition of glandular exudates of males and females was different both qualitatively and quantitatively.

Stoddart (1973) showed that the onset of sexual maturity in males of the rodent species *Apodemus flavicollis* was associated with a sudden increase in chemical complexity of the secretion produced by the caudal organ. Although he did not attempt to measure any rise in dominance, a rise is inferred by the field observations of newly mature young mice apparently having established individual ranges amongst not only their peers but also their parents' generation (Stoddart, unpublished).

The amount of secretion produced is known to depend, in one

rodent species at least, upon the amount of social experience gained by a test subject. Beauchamp (1974) examined the perineal scent gland in isolated male guinea pigs, and found that sebum production increased more than two-fold after social grouping for six weeks in dominant males and nearly two-fold in less dominant males.

Thus the picture that emerges is one of increased scent production, increased marking behaviour, and qualitative change in odour associated with a rise in social dominance and everything else that high status confers. It seems, from a consideration of response behaviour towards dominant odour, that dominance is treated differentially in different species. Jones and Nowell (1973) have shown that subordinate male house mice shy away from a part of an arena over which a dominant male has urinated. The precise site of manufacture of this aversive factor is not known. A similar response is seen in rats. When presented with cardboard cylinders which had previously housed for one hour either a dominant or a submissive strange male, rats of all dominance ranks show a significant increased interest in the cylinder tainted by a submissive male (Kramer *et al.*, 1969). The picture is complicated by the observation that no preference was shown by subjects taken from unstable hierarchies. The odorous factor would appear to be more effective on males living in stable hierarchies than on those from unstable hierarchies. Male marmosets, on the other hand, investigate more and spend more time by perches which have previously been marked by dominant males than by those previously marked by subordinates (Epple, 1973). The same applies to female subjects.

One instance has been reported in which the scent secretions of the dominant one or two males regulate much of the activity of the group. In the marsupial sugar glider (*Petaurus breviceps*), Schultze-Westrum (1965) reports from both field and laboratory studies that marking and territory patrolling behaviour, mating behaviour, aggressiveness, and even body weights are depressed by the presence of the community odour broadcast by the dominants. Secretions from the frontal and sternal glands, as well as the urine, are used for this purpose but it is not known which pieces of information are transmitted by which secretion. Histological observations of the frontal and sternal glands reveal that scent production was also depressed by the presence of community odour. All these behaviours and glandular functions quickly increased upon the removal of the dominant male but reverted to their former depressed state upon his reintroduction. *Petaurus* community odour can be seen to function much like the developmental depressive substances produced by the social hymenoptera and termites.

The demarcation and maintenance of territory

Most species of mammals relate in some way to a particular piece of land. A tract of land may be actively defended by one animal or a group against intruders, in which case it is usually called a territory, or a tract may be shared to a greater or lesser extent with others, in which case it is called an individual or a home range. Even migratory species maintain spatial proximity to others, although no land tenure may be involved. Odours appear to have two distinct uses in the territory and range context. The first is in demarcation of the range boundary and the second is for reassurance of the occupier when he is within his own domain.

There are many examples known of territorial boundaries being demarcated with odorous secretions. Frequently such scent marks are made from urine and/or faeces (e.g. in *Hippopotamus*) but equally frequently secretions from specialized sebaceous glands are used. Regrettably little is known about the effect of boundary marks on would-be intruders, but what evidence there is suggests that most marks serve to deter intruders. In this way the odours act to ensure the resources within a territory remain the exclusive right of the occupier and afford him and his family the best chances of survival. Mech (1970) reports that stray wolves (*Lupus lupus*) are visibly deterred by the presence of urine or faeces marks left by residents and frequently move away from marking places. Artificial spreading of castoreum from beavers (*Castor canadensis*) within an established territory elicits from the occupier intense investigatory and threatening behaviour. Aleksiuk (1968) has remarked that live trapping studies of this species indicate that transient beavers seldom enter the territories of established males. Since it is likely they are scared off by the odour emanating from the scent mounds, it can be seen that the population is effectively saved from growing too large, and thereby damaging its food resources, by the inhibitory effect of odour. As beaver populations over-produce young to a great extent each breeding season, an effective means for prevention of undue settlement is of great necessity. Lederer (1950) has identified 45 chemical constituents of castoreum. Many of them normally occur in mammalian urine and are deposited in the castor gland without conjugation along with several others which occur in the food. Which one or more of these compounds is responsible for the elicitation of avoidance reaction is not known.

The second use of odour in the territorial context is in reassuring the occupier when he is within his domain. This is definitely known to occur in rabbits (Mykytowycz, 1974) but is probably of more widespread occurrence. Rabbits mark prominent objects within the territory—stones, branches, and even tussocks of grass. Such familiar smelling landmarks serve to bolster the self-confidence of the occupant who is better able to win in any eventual conflict situation. The odour saturation thus aids survival and enhances individual longevity. When outside the home territory rabbits seldom defaecate or urinate and they maintain a nervous, raised gait the likes of which is not seen in the home territory.

Boundary marks are of use in species which maintain discrete territories or fixed individual ranges. Any one mark may influence just a handful of neighbours and strangers. In many species communal latrine sites, to which very many individuals repair for defaecation and urination, serve a function more akin to general information centres. Mykytowycz and Gambale (1969) and Mykytowycz and Hesterman (1970) report that dunghills within rabbit colonies serve an important social role both within and without the colony. Dunghills are frequented by all males and some females, and the daily addition of faecal pellets is not equal on all sites. From observational studies under artificial conditions it is clear that dunghills do not repel intruders; rather they impress upon strangers that they are entering an unfamiliar area and are subject to aggressive treatment. In this way the exclusiveness of the group territory is maintained. Young colony members are uninterested in dunghills, the greatest interest being shown by dominant males (Mykytowycz and Hesterman, 1970).

Wynne-Edwards (1962) mentions dunghills of the hyaena (*C. crocuta*) which may cover more than 1000 m^2. Hippopotamuses utilize large numbers of latrines on the river banks by their foraging trails. Eaton (1970) has indicated that cheetahs (*Acinonyx jubatus*) urinate on prominent stones or bushes within a communal hunting area shared by many. Practically nothing is known of what information is available in odour marks left at communal latrines, though it can be inferred that none is highly aversive. Whether communal latrines play a part in epideictic displays as claimed by Wynne-Edwards (1962) remains to be elucidated by experimental manipulation and behavioural observation.

The transmission of danger and warning signals

Odours are not well suited to the rapid transmission of their message since the speed at which they travel depends upon the diffusion coefficient of the substance in air and the speed of movement of the air. Nevertheless, there is some *prima facie* evidence that release of odour may be associated with the sensing of danger, though whether the odour acts as anything more than a reinforcing agent to acoustic and visual signal transmission is yet unresolved. If it plays any role at all it will influence and enhance population survival by increasing individual longevity. Many species of the larger mammals bear tufts and flashes of contrasting coloured hair which are capable of being erected. Smyth (1970) points out that these (usually) white patches may serve to attract predators and draw an attack while the intended victim is still within its own range and in sensory contact with the predator. This may be one of the functions of the white flash but an investigation of the cutaneous glands associated with it suggests another. The springbok (*Antidorcas marsupialis*) bears a distinct pouch from 15 cm to 23 cm long lying from the base of the tail forward along the spine. The pouch is lined with secretory epithelium producing a yellow, sticky secretion. The lateral walls of the pouch bear long, springy, white hairs. When the animal is at rest the pouch is closed and the hairs are not exposed. When disturbed or alarmed the pouch opens and pilo-erector muscles expose the long white hairs. It is reasonable to assume that the adhering secretion rapidly evaporates from the surface of the hairs (Schaffer, 1940). In the rock cavy (*Dendrohyrax dorsalis*) a dorsal white flash overlies a yellow glandular patch measuring 22 mm × 9 mm. When alarmed the hyraxes flash their white patches and presumably cause some of the underlying secretion to be broadcast. In the mara, or Patagonian hare (*Dolichotis patagonum*) there are a pair of supra-anal glands which are overlain by a broad band of erectile white hairs. This patch is everted whenever the animals are alarmed. The pronghorn antelope (*Antilocapra americana*) displays twin white patches lateral, and slightly dorsal, to the anus. This species bears a single supracaudal gland which produces a pungent oily secretion. Erection of the twin patches causes some of the secretion to be picked up by the long hairs and, presumably, dispersed. Regrettably no experimental work appears to have been performed on this odour system and so the interpretations presented here must remain conjectural. The odours may play a role in other behaviours in all species since piloerection is not restricted to alarm situations.

Conclusions

It can be seen that the field of olfactory investigation with respect to population biology is still very much underdeveloped. Yet in a few studies, notably that currently being undertaken by Mykytowycz and his colleagues on the rabbit, the importance of odours in the social biology of the species is rated very high. There are scarcely any aspects of the species' daily life in which odour production or reception does not play a part. There seems no reason to doubt that odours are equally important to most, if not all, species of mammals. The requirement now is for studies to be directed towards an understanding of precisely what effect at the population level the experimental introduction or deletion has of a particular odour, or complex of odours. Among the questions to be asked are whether extrinsically administered odorous preparations can bring about changed reproduction through some change in the quantity or quality of social interaction, and whether dispersion and dispersal of the population can be modified by odour cues. One can speculate that such work, being in an area of fundamental importance to the social wellbeing of the species, will find a useful application in the field of pest control and, with the problems of toxic pesticides widely known, it is surprising that basic olfactory studies on pest species' populations have not received the emphasis and support they deserve.

References

Aleksiuk, M. (1968) Scent-mound communication, territoriality, and population regulation in beaver (*Castor canadensis* Kuhl). *Journal of Mammalogy*, **49**, 759–762.

Beauchamp, G. K. (1974) The perineal scent gland and social dominance in the male guinea pig. *Physiology and Behaviour*, **13**, 669–673.

Bruce, H. M. (1960) A block to pregnancy in the mouse caused by proximity of strange males. *Journal of Reproduction and Fertility*, **1**, 96–103.

Drikamer, L. C., Vandenbergh, J. G., and Colby, D. R. (1973) Predictors of dominance in the male golden hamster (*Mesocricetus auratus*). *Animal Behaviour*, **21**, 557–563.

Eaton, R. (1970) Group interaction, spacing and territoriality in cheetahs. *Zietschrift für Tierpsychologie*, **27**, 481–491.

Epple, G. (1973) The röle of pheromones in the social communication of marmoset monkeys (Callithricidae). *Journal of Reproduction and Fertility*, Supplement 19, 447–454.

Epple, G. (1974) Primate pheromones. In *Pheromones*, ed. Birch, M.C. pp. 366–385. Amsterdam: North-Holland Publishing Company.

Goodrich, B. S. and Mykytowycz, R. (1972) Individual and sex differences in the chemical composition of pheromone-like substances from the skin glands of the rabbit, *Oryctolagus cuniculus*. *Journal of Mammalogy*, **53**, 540–548.

Hesterman, E. R. and Mykytowycz, R. (1968) Some observations on the odours of anal gland secretions from the rabbit, *Oryctolagus cuniculus* (L). CSIRO *Wildlife Research*, **13**, 71–81.

von Holst, D. and Buergel-Goodwin, U. (1975) Chinning by male *Tupaia belangeri*: the effects of scent marks of conspecifics and of other species. *Journal of Comparative Physiology*, **103**, 153–171.

Johnson, R. P. (1973) Scent marking in mammals. *Animal Behaviour*, **21**, 521–535.

Jones, R. B. and Nowell, N. W. (1973) Aversive and aggression-promoting properties of urine from dominant and subordinate male mice. *Animal Learning and Behaviour*, **1**, 207–210.

Krames, L., Carr, W. J., and Bergman, B. (1969) A pheromone associated with social dominance among male rats. *Psychonomic Science*, **16**, 11–12.

van der Lee, S. and Boot, L. M. (1956) Spontaneous pseudopregnancy in mice. II. *Acta Physiologica Pharmacologica Neerlandica*,**5**, 213–214.

Lederer, E. (1950) Odeur et parfums des animaux. *Fortschritte der Chemie und Organische Naturstoffe*, **6**, 87–153.

Leon, M. and Moltz, H. (1972) The development of the pheromonal bond in the albino rat. *Physiology and Behavior*, **8**, 683–686.

Martan, J. (1962) Effect of the castration and androgen replacement on the supracaudal gland of the male guinea pig. *Journal of Morphology*, **110**, 285–297.

Mech, L. D. (1970) *The Wolf: the Ecology and Behaviour of an Endangered Species*. New York: Natural History Press.

Mykytowycz, R. (1965) Further observations on the territorial function and histology of the submandibular cutaneous (chin) glands in the rabbit, *Oryctolagus cuniculus* (L). *Animal Behaviour*, **13**, 400–412.

Mykytowycz, R. (1974) Odor in the spacing behavior of mammals. In *Pheromones*, ed. Birch, M. C. pp. 327–343. Amsterdam: North Holland Publishing Company.

Mykytowycz, K. and Dudzinski, M. L. (1966). A study of the weight of odoriferous and other glands in relation to social status and degree of sexual activity in the wild rabbit, *Oryctolagus cuniculus* (L.). CSIRO *Wildlife Research*, **11**, 31–47.

Mykytowycz, R. and Gambale, S. (1969) The distribution of dung-hills and the behaviour of free-living wild rabbits, *Oryctolagus cuniculus* (L.), on them. *Forma et Functio*, **1**, 333–349.

Mykytowycz, R. and Hesterman, E. R. (1970) The behaviour of captive wild rabbits, *Oryctolagus cuniculus* (L), in response to strange dung-hills. *Forma et Functio*, **2**, 1–12.

Ralls, K. (1969) Scent marking in Maxwell's duiker, *Cephalophus maxwelli*. *American Zoologist*, **9**, 1071.

Ralls, K. (1971) Mammalian scent marking. *Science*, **171**, 443–449.

Schaffer, J. (1940) Die Hautdrüsenorgane der Säugetiere. Berlin: Urban and Schwarzenberg.

Schultze-Westrum, T. (1965) Innerartliche Verständigung durch Düfte beim Gleitbeutler, *Petaurus breviceps papuanus* Thomas (Marsupialia: Phalangeridae). *Zeitschrift für Vergleichende Physiologie*, **50**, 151–220.

Smyth, N. (1970) On the existence of 'pursuit invitation' signals in mammals. *American Naturalist*, **104**, 491–494.

Stoddart, D. M. (1972) The lateral scent organs of *Arvicola terrestris* (Rodentia: Microtinae). *Journal of Zoology, London*, **166**, 49–54.

Stoddart, D. M. (1973) Preliminary characterisation of the caudal organ secretion of *Apodemus flavicollis*. *Nature, London*, **246**, 501–503.

Stoddart, D. M. (1976) *Mammalian odour and pheromones*. Studies in Biology No. 73. London: Edward Arnold.

Stoddart, D. M. (1977) Two hypotheses supporting the social function of odorous secretions of some old world rodents. In *Chemical Signals in Vertebrates*, ed. Müller-Schwarze, D. Ch. 19, pp. 333–355. New York: Plenum.

Stoddart, D. M., Aplin, R. T., and Wood, M. J. (1975) Evidence for social difference in the flank-organ secretion of *Arvicola terrestris* (Rodentia: Microtinae). *Journal of Zoology, London*, **177**, 529–540.

Thiessen, D. D. (1968) The roots of territorial marking in the Mongolian gerbil: a problem of the species-common topography. *Behavior Research Methods and Instrumentation*, **1**, 70–76.

Vrtis, V. (1930) [Glandular organ on the flanks of the water rat, their development and changes during breeding season.] *Biologie Spisy vysoke Škola zvěrolék Brno*, **4**.

Whitten, W. K. (1959) Occurrence of anoestrus in mice caged in groups. *Journal of Endocrinology*, **18**, 102–107.

Wynne-Edwards, V. C. (1962) *Animal dispersion in relation to social behaviour*. Edinburgh: Oliver and Boyd.

Physiological mechanisms of stress

J. R. CLARKE

Department of Agricultural Science, University of Oxford, Oxford, England

Introduction

Stress may be considered as a damaging or threatening environmental factor which elicits hyperactivity of the adrenal cortex and the sympathetic nervous system. Plasma corticosteroid, and catecholamine released from the adrenal medulla and post-ganglionic sympathetic nerve endings, increase. It is, however, important to realize that corticosteroid output normally displays diurnal fluctuation, independent of obvious stress; plasma levels are highest in the early morning and are low in the evening in sheep, rat, and mouse (Sharp *et al.*, 1961; Haus and Halberg, 1970; Yates *et al.*, 1971; McNatty *et al.*, 1972; Ramaley, 1976). In man and sheep periodic changes in corticosteroids are superimposed, as fairly regular peaks, upon this circadian (24 hour) rhythm. Data from normal man, and from man living under a reversed rhythm of activity, suggest that the corticosteroid rhythm is endogenous, but that its phase is determined by the light/dark regime, by wakefulness and sleep, and/or by activity and inactivity. Stress produces corticosteroid levels in excess of circadian variation and improves the animal's capacity to adapt to the stress; for example the effects of corticosteroids and catecholamines separately or together upon carbohydrate metabolism and on the cardiovascular system may be considered beneficial, although not all the effects, particularly of corticosteroids, are advantageous. The activation and time course of these two systems by a stressor may be similar, so that a functional connection between them may be expected.

There are three principle corticosteroids secreted by eutherian mammals, cortisol (compound F), corticosterone (compound B), and aldosterone. The stress release of cortisol or corticosterone involves a neurohumoral activation of nerve cells in the basal hypothalamus or of nerve endings in the median eminence (Harris, 1955; Szentagothai *et al.*, 1968; Harris, 1972; Brodish, 1973); the release into the hypothalamo–hypophysial portal system of corticotrophin-releasing factor (CRF) induces secretion of adrenocorticotrophic hormone (ACTH) from the anterior pituitary gland; the ACTH stimulates synthesis and release of cortisol and/or corticosterone from the adrenal cortex. Aldosterone production is chiefly controlled by the renin–angiotensin system and to a much lesser extent by ACTH. The activities of the renin–angiotensin system and ACTH are not readily separable, however. Corticosteroids, released as a result of ACTH action, themselves regulate the output of ACTH, either by direct actions on the anterior pituitary gland, or indirectly by effects upon neurones in the forebrain. These general relationships are represented in figure 1. The importance of the C.N.S. and of corticosteroid feedback to the activity of the anterior pituitary-adrenocortical system is illustrated in the following way (Ganong, 1963). Unilateral adrenalectomy causes enlargement of the remaining adrenal, but this will not occur if a lesion is made in the median eminence. Two inferences may be drawn from such observations. Firstly, secretions (corticosteroids) from the adrenal gland directly or indirectly suppress ACTH release from the anterior pituitary gland. Secondly, central control of ACTH output is in the median eminence and/or higher centres of the brain.

The study of stress and its effects involves the following problems:

1. Can different sorts of stress be recognized and what are the sensory pathways to the hypothalamus?
2. How do the effects of stress override the basal regulation of ACTH and corticosteroid release, which involves negative feedback effects of corticosteroids on the hypothalamo–hypophysial system?
3. What are the relationships between the response to stress of the ACTH–adrenocortical system, and the catecholamine-releasing systems of the central and sympathetic nervous systems and of the adrenal medulla?
4. What effects do the increased corticosteroids and catecholamines have on other endocrine mechanisms such as those controlling reproduction?

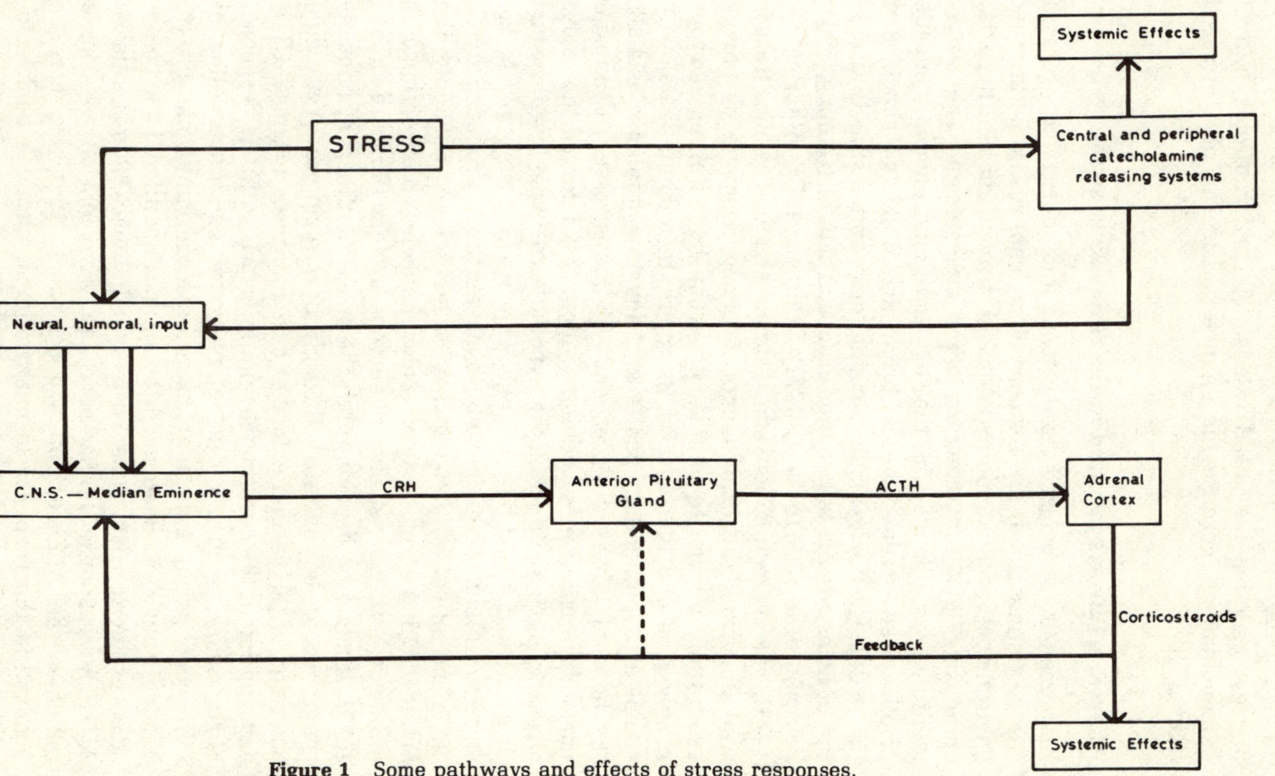

Figure 1 Some pathways and effects of stress responses.

5. How may the increased release of corticosteroids and of catecholamines increase the chance of survival of a stressed animal?

Afferent pathways involved in the response to stress

It is generally recognized that systemic and neurogenic stressors can be distinguished. The former involve changes in the internal environment, and include hypotension, hypoglycaemia, hypoxia, endotoxins, and the effect of ether anaesthesia. The latter, identified as changes in the external environment, include emotional stress, loud noises, bright lights, electric shock, and tissue damage (for example burning, nociceptive stimuli, breaking of limbs, or the application of a tourniquet) (Ganong, 1963; Feldman *et al.*, 1970, 1971, 1972; Feldman, 1973; Allen *et al.*, 1973). The distinction between systemic and neurogenic stress is well illustrated by experiments involving complete or partial separation of the hypothalamus from its afferent nervous connections and therefore of its input. In animals (rats) with complete hypothalamic deafferentation (hypothalamic island), the adrenal cortex still responds to the stresses of anoxia, immobilization, or ether administration. It is inferred that these stressors have an effect on the hypothalamo–hypophysial system through some blood-borne agent. This would also apply to the stressful effects of endotoxins, such as those produced by *E. coli*. The effect of hypotension is less clear. Certainly the enhanced output of corticosteroids following haemorrhage is influenced by the peripheral nervous system (Redgate *et al.*, 1973). It might be expected that the stressful effects of haemorrhage would be initiated through baroreceptors. Yet while occlusion of the carotid artery of cats increases plasma ACTH, denervation of the carotid does not interfere with this response. Furthermore, section of the spinal cord and denervation of baroreceptors in cats that have been adrenalectomized (to eliminate any catecholamine effects), does not alter the hypotensive response. Also, cats with hypothalamic islands release ACTH in response to haemorrhage. Hypotension of course activates the renin–angiotensin system so that this may be involved in the response to haemorrhage, although nephrectomized but otherwise intact cats still release ACTH after haemorrhage. However, the ACTH response is considerably reduced by combining nephrectomy with denervation of the baroreceptors. When the superior cervical ganglia is also removed, the ACTH response is completely eliminated. Indeed, the sympathetic nerve supply to the head seems (at least in cats) to play an appreciable role in the control of ACTH release, since stimulation of the sympathetic supply directly, or by the infusion of bradykinin into the arterial supply of the superior cervical ganglion, increases ACTH output (see also Allen *et al.*, 1973). However, in the case of the direct stimulation of the cranial sympathetic supply, it has to be remembered that visceral afferent nerves accompany the efferents. Following nerve section, stimulation on the distal side of the cut does not enhance ACTH output, whereas stimulation of the proximal side causes an ACTH response. Thus there is a neurogenic component in adrenocortical response to haemorrhage. The overall response to the stress of haemorrhage seems to involve: (*a*) sino-aortic baroreceptors, (*b*) autonomic innervation of the head, (*c*) the renin–angiotensin system (see also Gann and Cryer, 1973). The systemic element in this response to the stress of haemorrhage is strongly suggested by the observation that infusion of angiotensin II into the arterial supply of the head causes an increase in the release of ACTH (or cortisol) and, at least in dogs, this has been shown to occur in those which have an isolated median eminence—anterior pituitary system (Redgate *et al.*, 1973; Gann and Cryer, 1973).

The importance of afferent pathways in stress responses becomes apparent from simple experiments. If the limb of a mammal is burnt, the ACTH output increases, an effect that still occurs when the circulation of the limb is occluded provided the afferent (sensory) nerves are intact. The enhanced release of ACTH will not however take place if the sensory nerves of the limb are severed, even though the circulation is undisturbed (Ganong, 1963). Other peripheral traumatic stressors, such as leg fracture or the application of a tourniquet to a hind limb, increase ACTH via afferent pathways ascending in the contralateral spinal cord, through the pons, dorsal mesencephalon, and median forebrain bundle, to enter the basal anterior quadrant of the hypothalamus on that side (Gibbs, 1969; Allen *et al.*, 1973).

Systemic and neurogenic stressors may also be distinguished in experiments involving deafferentation of the hypothalamus. Anterior deafferentation of the hypothalamus in rats does not alter responses to ether, acoustic, or photic stress, while antero-lateral or postero-lateral deafferentations have little influence on the response to ether stress but interfere with the response to photic or acoustic stress. These lesions all affect the median forebrain bundle connecting midbrain and limbic systems. Such pathways may be expected to transmit acoustic stimuli from the lower brain stem. The photic stimuli do not appear to have a direct route to the hypothalamus, being conveyed by the inferior fasiculus of the accessory optic tract, lying in the median forebrain bundle, ending in the medial terminal nucleus within the ventral midbrain tegmentum. This nucleus is lateral to the mammillary peduncle, which itself contributes to the median forebrain bundle. Thus the route taken by the photic neurogenic stress is: eye ⟶ optic chiasma ⟶ accessory optic tract ⟶ medial terminal nucleus ⟶ mammillary peduncle ⟶ median forebrain bundle ⟶ hypothalamus ⟶ anterior pituitary ⟶ adrenal cortex (Feldman *et al.*, 1971, 1972; Feldman, 1973). Quite simply, considering the stress of auditory or optical stimulation, and of ether, anoxia, and immobilization, the response to the former requires a neural input to the medial basal hypothalamus, whereas that to the latter involves a humoral activation of the hypothalamus (Feldman *et al.*, 1970). Later it will be necessary to consider in some detail the significance of catecholamines acting upon the hypothalamus for the response of animals to stress. But dealing now with the neural pathways involved in the initiation of the response to stress, the importance of noradrenergic and (to a lesser extent) serotinergic pathways should be recognized. Two noradrenergic pathways arise from cell groups in the hind brain (medulla and pons): a ventral bundle which innervates the hypothalamus and pre-optic area; and a dorsal bundle, arising from the cells of the locus coeruleus, which innervates the cerebral cortex, hippocampus, and cerebellum (Ungerstedt, 1974). Lesions in the ventral bundle (for example, those produced by local application of 6-hydroxydopamine) increase the response to ether stress although this effect disappears a few days later. Lesion of the dorsal bundle by the same method does not affect the response to stress, although it does influence conditioned avoidance behaviour: the fixation of a new conditioned avoidance response is decreased in such lesioned and adrenalectomized rats. This could indirectly have consequences for at least certain sorts of neurogenic (emotional) stress. Lesion of the serotinergic pathways between the raphe nucleus of the hind-brain (pons) and the hypothalamus, through the local infusion of 5, 6-hydroxytryptamine, increases *basal* (as opposed to stress-enhanced) plasma corticosteroid levels shortly after the lesion is made, but this response also wears off after several days (Fuxe *et al.*, 1973).

There does, then, seem to be some understanding of the pathways by which stresses of different sorts are relayed to the hypothalamic–anterior pituitary–adrenocortical system. Depending on the pathways involved, a simple, but probably not absolute, distinction can be drawn between neurogenic and systemic stresses. It is also important to realize that because of insufficient understanding of the *grading* of stress it is impossible to decide whether or not the supposed differentiation into types is due in some measure to differences in stimulus strength (see Ganong, 1963). Another type of distinction has been made between stress which is resistant, and stress which is sensitive, to the feedback effects of corticosteroids (Sato *et al.*, 1975).

The feedback control of hypothalamic and adenohypophysial activity

It is generally accepted that corticosteroids regulate the output of ACTH by inhibiting its release. This could be inferred from (*a*) the compensatory adrenal hypertrophy which follows unilateral adrenalectomy, (*b*) the rise in plasma ACTH after bilateral adrenalectomy, (*c*) the capacity of corticosterone or dexamethasone to reduce plasma ACTH levels in bilaterally adrenalectomized animals, and (*d*) the fall in the ACTH levels of intact animals which occurs as endogenous corticosterone levels rise (Rees *et al.*, 1971; Dallman *et al.*, 1972; Dallman and Jones, 1973a,b; Gann and Cryer, 1973). Such negative feedback control of ACTH release ensures fairly constant basal levels of corticosteroids.

Is the negative feedback action of corticosteroids directly

upon the anterior pituitary, or at a site in the central nervous system, such as the hypothalamus or median eminence? Experimental evidence supports both possibilities, although results increasingly emphasize the importance of the C.N.S. In support of the idea of direct feedback on the pituitary, it has been found that the absence of the forebrain and median eminence does not prevent dexamethasone inhibition of the endogenous release of corticosteroids. Very high doses of corticosteroids will depress adrenocortical function in hypophysectomized rats with pituitaries transplanted under the kidney capsule: it seems unlikely that these transplanted glands would have been under the remote control of the C.N.S. Incubating pituitary glands with extracts of stalk-median eminence tissue will cause the release of ACTH into the incubation medium. However the output of ACTH is reduced if the pituitaries are first incubated in a medium containing dexamethasone. These *in vitro* results suggest that corticosteroids can inhibit ACTH release by blocking the response of the pituitary to corticotrophin-releasing factor (Kendall, 1971). On the other hand, the profiles of CRF, ACTH, and corticosteroid release in stressed rats, and the temporal relationships in the output of these hormones, suggest strongly that feedback control of ACTH is at the level of the hypothalamus (Sato *et al.*, 1975). Perhaps there are several interdependent sites of feedback action, including the anterior pituitary, the hypothalamus, and other parts of the C.N.S. (see below, pp. 136-140). Their relative importance may change from the circumstances of maintenance of basal levels of corticosteroids to those of stress when greatly enhanced levels are sustained for some time.

How are the elevated titres of corticosteroids maintained under conditions of stress? If the relationships between the hypothalamo–hypophysial system and the adrenal cortex were the same in the stressed as in the unstressed state, then no sooner would the titre in plasma of corticosteroids rise than negative feedback effects would reduce ACTH output, in turn lowering corticosteroid levels. This does not happen. Levels of corticosteroids, and in some cases also of ACTH and CRF, have been shown to be considerably elevated for several hours, both during and after cessation of the stress (Dallman and Jones, 1973a,b; Sato *et al.*, 1975). Apparently stress alters for some little time central nervous mechanisms to lessen negative feedback effects. Evidence for this phenomenon may now be considered (see also Dallman and Jones, 1973a,b).

It has been found that when rats are exposed to two stressors (electric shock) at intervals of from 1 to 24 hours, the response to the second stressor (in terms of plasma corticosterone titres) is of about the same magnitude as to the first. Physical restraint of rats is a very powerful stressor, and physical restraint for 90 minutes increases plasma corticosterone levels, an effect sustained for some time after the cessation of the stress. Nevertheless, the adrenocortical response to a more trivial stress (injection) four hours after the restraint has ended is quite as large in such rats which have already been severely stressed as in others not subject to prior stress. Further evidence has been obtained suggesting that an initial stressor leaves some trace on central mechanisms which influences the responsiveness of the system. Two groups of rats had their plasma ACTH and corticosterone levels raised to about the same levels, in one group by stress and in the other by injection of ACTH or corticosterone. Some hours later *both* groups were stressed. The rats which had already been stressed once (and whose corticosterone levels had at that time been *actively* raised) responded more strongly to their second stressor than those whose corticosterone levels had been passively increased by the administration of hormones and had then been exposed to the same (but for them their first) stressor.

In addition to an initial stressor effect on central nervous mechanisms that affects reaction to a second stress, it seems that there are two sorts of feedback mechanism. One is the fast, rate-sensitive feedback, the other the slow, level-sensitive feedback (Dallman *et al.*, 1972; Yates and Maran, 1974; Sato *et al.*, 1975). The fast feedback is sensitive (in a manner not yet understood) to the rate of rise in plasma corticosteroids, as exemplified by the response of rats to sham adrenalectomy (as a stressor). There is a rapid rise in plasma ACTH, which peaks 2.5 minutes after the stress. The rise in plasma corticosterone follows this, peaking at five minutes. As the corticosterone rises, so that ACTH titres fall. In adrenalectomized animals there is, immediately following the operation, a rapid, steep rise in plasma ACTH. This can be controlled and caused to fall by injection of corticosterone at a dose which gives plasma corticosterone levels very similar to those of intact, stressed rats. This fast feedback action of corticosterone is modified by a natural secretory product of the rat adrenal cortex, 18-hydroxy-deoxycorticosterone. It apparently *enhances* the response of the anterior pituitary–adrenocortical system to ether stress when it is injected four hours beforehand, possibly by reducing the corticosterone fast feedback. If, ten minutes before a rat is stressed, it is treated with *corticosterone*, the ACTH output in response to the stress is reduced. However if 18-hydroxydeoxy-corticosterone is administered some minutes before the rat is injected with corticosterone, the inhibitory effects of the corticosterone upon ACTH output are reduced (Tiptaft and Jones, 1976).

The slow, proportional feedback was suggested by the following experiment. One group of rats was bilaterally adrenalectomized, and another sham operated, these procedures being regarded as a first stress. One, two, or 24 hours afterwards, rats in both groups were subjected to a second stressor, laparotomy. Within each group, the size and duration of the rise in plasma ACTH was the same at the three intervals following the initial stress. However ACTH reached much higher levels in the adrenalectomized rats than in the sham-operated animals. The corticosterone secreted in response to the first stress somehow inhibited for up to 24 hours ACTH release in response to the second stress.

In certain circumstances another mechanism permits the steroid sensitive feedback elements of the brain stem to be overridden or bypassed, to enhance ACTH output. In dogs, for example (Gann and Cryer, 1973), large haemorrhage produces cortisol release. The response depends on baroreceptors and on the kidney, as experiments involving vagal and carotid denervation and nephrectomy have shown. Angiotensin II is an important component and acts on the median eminence to provoke ACTH release. This mechanism is thus independent of the steroid sensitive feedback system, although a second bypass may exist.

The hypothalamic corticotrophin-releasing factor secreting neurones are inhibited by noradrenergic receptors (see below). In the studies of Gann and Cryer (1973) four types of dog were severely haemorrhaged: (*a*) normal dogs, (*b*) dogs treated with dexamethasone, (*c*) dogs given iproniazid (a monoamine oxidase inhibitor, which potentiates actions of noradrenergic pathways), and (*d*) dogs given dexamethasone and iproniazid. There was significant suppression of ACTH release in dogs given dexamethasone, but not in those given iproniazid. However, a more complete suppression of ACTH release was achieved after both dexamethasone and iproniazid. Thus baroreceptor pathways project onto the noradrenergic pathways which otherwise tend to suppress ACTH output. From these and other experiments on the effects of haemorrhage in dogs it is suggested that there is: (*a*) a low threshold component suppressable by dexamethasone regulating ACTH output, (*b*) a moderate threshold component (within the median eminence) responding to the renin–angiotensin system; (*c*) a high threshold bypass, coming into action with reduced carotid baroreceptor stimulation, affecting the median eminence.

The feedback mechanisms discussed above arise from measurements of blood hormonal levels following a variety of experimental procedures. However direct manipulation of the hypothalamus and of parts of the adjacent limbic system have also been employed.

The vascular connections between the median eminence and the anterior pituitary, and the neuronal links between the hypothalamus and the median eminence, clearly suggest that if the feedback regulation of ACTH reflects C.N.S. activity, an obvious site of action is within the hypothalamus and median eminence, as discussed above. Support comes from observed effects of hypothalamic stimulation or of hypothalamic lesions which, when appropriately localized, cause on the one hand the release of adrenocorticosteroids and on the other, block the response to stress (Harris, 1955, 1972; Szentagothai *et al.*, 1968). Since bilateral adrenalectomy of female rats causes significant neuronal nuclear enlargement in parts of the ventromedial nucleus of the hypothalamus, Palkovits and Stark (1972) suggested that most of the corticosteroid sensitive neurones are in circumscribed parts of the ventromedial and arcuate nuclei. However it is conceivable that, although these neuronal changes may well be effects of adrenalectomy, they are 'downstream' from the site of action of corticosteroids.

Electrophysiological studies implicate the anterior hypothalamus and median eminence in corticosteroidal feedback actions. Intraperitoneal injection of cortisol appreciably reduces neuronal activity in these regions which otherwise follows stressful photic and acoustic stimuli or stimulation of the sciatic nerve (Feldman, 1973).

It is possible to incubate hypothalami and to estimate the amount of corticotrophin releasing factor discharged into the incubating fluid, as well as measuring it in the hypothalamic tissue. The hypothalami of intact rats, those which had been adrenalectomized or adrenalectomized and treated with corticosteroids, have been investigated for their capacity, *in vitro*, to release CRF (Hillhouse and Jones, 1976). Thus adrenalectomy increases the release of hypothalamic CRF in the following 4 to 24 hours, and it becomes very marked by seven days. This enhancement can be sustained for three months following adrenalectomy. Acetyl choline and 5-hydroxytryptamine (5-HT = serotonin) *in vitro* cause an increase in CRF output from the hypothalami taken from adrenalectomized rats, without reducing hypothalamic CRF content. However acetyl choline is inactive on hypothalami from adrenalectomized rats treated with corticosterone. Thus it is suggested that: (*a*) corticosteroids have a negative feedback action on hypothalamic CRF production; in these experiments the effects *in situ* on the hypothalamus occurred some hours after corticosteroid administration, and so would appear to have involved a slow feedback mechanism; (*b*) two excitatory neurotransmitters, acetylcholine and 5-hydroxytryptamine, cause CRF release and synthesis, the two steps perhaps being coupled.

Other work suggests that 5-hydroxytryptamine is involved in the regulation of the CRF–ACTH–corticosteroids system. In rats, stress causes rapid release of 5-hydroxytryptamine from axonal terminals in the hypothalamus, the cell bodies of which lie in the raphe nucleus, and an increase in tryptophan hydroxylase activity in the mid-brain; tryptophan hydroxylase is the rate-limiting enzyme in 5-HT biosynthesis. Furthermore, bilateral adrenalectomy prevents the increase in this enzyme which follows stress, while treatment with corticosterone increases the enzymic activity (Palkovits *et al.*, 1976; Azmita and McEwen, 1974). Such findings have been taken as evidence for a reciprocal relationship between the serotinergic and the pituitary–adrenal systems, the former inhibiting the latter (Telegdy and Vermes, 1976).

There are undoubtedly extrahypothalamic centres which influence ACTH output; in particular the hippocampus seems to inhibit CRF–ACTH–corticosteroid output. Electrical stimulation of the hippocampus decreases corticosteroid levels and corticosteroid implants in this region raise them (Michel, 1974). Dexamethasone infused into the dorsal hippocampus decreases the neuronal activity (Michel, 1974; Segal, 1976). Such actions of dexamethasone and corticosteroids implanted in the hippocampus may represent bypass mechanisms of the negative feedback, an override used in the response to stress.

The discovery that the hippocampus avidly takes up corticosteroids is obviously very relevant. Hippocampal sites are saturated by corticosterone levels such as are found in normal rats, so that hippocampal uptake is only readily demonstrated in adrenalectomized animals (McEwan *et al.*, 1969). A corticosterone-binding protein from brain cytosol has been identified and is present in highest concentrations in the hippocampus. It appears to be different from serum corticosterone-binding protein, having a high affinity and limited capacity for corticosterone. Furthermore, oestradiol, testosterone, and 11-dehydrocorticosterone do not compete with corticosterone for the sites on the binding protein. Thus the receptor appears stereospecific for a naturally occurring corticosteroid (McEwan *et al.*, 1969; Stevens *et al.*, 1971; Grosser *et al.*, 1973; McEwan and Wallach, 1973; McEwan *et al.*, 1974).

Another system exercises control over the hippocampus in a fashion which may be relevant to the stress response. The locus coeruleus (see above, p. 132) sends noradrenergic pathways to the hippocampus. Stimulation of the locus coeruleus, iontophoretic application to hippocampal pyramidal cells of noradrenaline or D.M.I. (which prevents the reuptake of noradrenaline), inhibits the activity of these cells, while a β-adrenergic antagonist (NJ-1999) blocks the inhibitory action of noradrenaline, and 6-hydroxydopamine given intracisternally causes hippocampal cells to fire faster (Segal and Bloom, 1974a,b). The capacity of noradrenaline to inhibit a system which tends to restrain ACTH output could be regarded as an appropriate adaptation, since in stress there is widespread activation of noradrenergic pathways (see below): the inhibitory effect of noradrenaline possibly released during stress from nerve endings in the hippocampus could promote the release of CRF and thus of ACTH. However, ACTH applied iontophoretically to the hippocampus *increases* the activity of some of its neurones. Perhaps ACTH is able in the intact animal to reach the hippocampus by a vascular supply akin to the short feedback loops between the anterior pituitary and the hypothalamus (Szentagothai *et al.*, 1968). Its effect there may be to modify those of noradrenaline (Segal, 1976).

In addition to the hippocampus, the amygdala participates in the control of the CRF–ACTH–adrenocortical system. In freely behaving conscious cats stimulation of the amygdala enhanced ACTH release when plasma levels of cortisol and corticosterone were low, and inhibited ACTH output when titres of these steroids were high. Thus there may be some negative feedback control at the level of an amygdaloid nuclei (Matheson *et al.*, 1971).

Catecholamines and the secretion/release of ACTH

The abundance of aminergic (and serotinergic) fibres in the hypothalamus and hippocampus, the demonstration by histochemical methods of catecholamines in hypothalamic tissue (Fuxe and Hökfelt, 1969; Hökfelt and Fuxe, 1972: Fuxe *et al.*, 1973), and the considerable release of noradrenaline from both post-ganglionic sympathetic nerve terminals and the adrenal medulla at least at the initial stages of stress (Callingham, 1975; Lewis, 1975) make it plausible that catecholamines are in some way involved in regulating ACTH release. Stress causes increased turnover of noradrenaline in the brain stem and hypothalamus (Gordon *et al.*, 1966; Stone, 1973; Thierry *et al.*, 1968, 1970), and at least for the medulla oblongata the response to cold stress is associated with induction of tyrosine hydroxylase, as revealed by increased *in vitro* tyrosine hydroxylase activity (tyrosine hydroxylase being the rate-limiting enzyme for noradrenaline biosynthesis). There is no significant induction of tyrosine hydroxylase in the hypothalamus. However, catecholamine nerve endings in the hypothalamus have cell bodies in the medulla, so the tyrosine hydroxylase induction in the medulla could affect catecholamine activity in the hypothalamus (Thoenen, 1970). Furthermore, stressing rats in the last few days of pregnancy causes their offspring, when 40–45 days old, to have higher turnover of noradrenaline in the telencephalon, diencephalon, and mesencephalon than offspring of unstressed females (Huttunen, 1971).

In contrast to the effects of noradrenaline at the *hippocampal* level on the CRF–ACTH system (see above, pp. 139–140), noradrenergic pathways are also inhibitory through their action on the *hypothalamus* as a result of noradrenaline α-receptors on the CRF neurones. Furthermore, notwithstanding evidence quoted earlier (p. 138) that 5-hydroxytryptamine may promote CRF release, there are reasons for supposing that 5-HT also inhibits the CRF–ACTH system. From a functional viewpoint, these conclusions are surprising, since the considerable increase in sympathetic activity under conditions of stress then might be expected to reduce the CRF–ACTH response. Recent investigations may have resolved this conflict (see later pp. 143–144). The release of 17-hydroxycorticosteroids in dogs after stimulation of the femoral nerve, the caudal medulla oblongata, the bundle of Schütz or the mamillary peduncle, is blocked by α-ethyltryptamine, a monoamine oxidase inhibitor which potentiates the effects of norandrenaline. However, if the ventral hypothalamus or median eminence of such dogs is directly stimulated, the output of 17-hydroxycorticosteroids increases, and α-ethyltryptamine does not reduce ACTH release in dogs with hypothalamic islands (Ganong *et al.*, 1965). Furthermore, it has been found in dogs that sympathomimetic drugs can inhibit the stress-induced ACTH secretion. L-Dopa will have this effect when drugs are given systemically, whereas noradrenaline and dopamine are only active when given into the third ventricle. L-Dopa crosses the blood–brain barrier, but noradrenaline and dopamine do not. It seems, then, that the catecholamines act to inhibit the CRF–ACTH system at a site *inside* the blood–brain barrier; this excludes the median eminence as a site of inhibition. Since phenoxybenzamine (an α-adrenergic antagonist) prevents, when infused into a brain ventricle, the block of ACTH release by L-Dopa, it is considered that this inhibitory system involves α rather than β receptors on CRF neurones. Further evidence consistent with this has come from experiments using clonidine, an α-agonist. In these circumstances stress release of ACTH is prevented. However if phenoxybenzamine is given intraventricularly, clonidine fails to have this effect.

Thus in the dog, a central aminergic system inhibits the CRF response to stress. The site of action is considered to be the hypothalamus. It involves α-adrenergic receptors on CRF neurones inside the blood–brain barrier. The mediator is noradrenaline rather than dopamine, and the noradrenergic neurones end on cell bodies and/or dendrites of putative CRF secreting neurones (Ganong *et al.*, 1976).

If catecholamines inhibit the release of CRF and thereby ACTH and corticosteroids, then depletion of hypothalamic noradrenaline ought to affect, for example, plasma corticosteroid levels. Guanethedine and 6-hydroxydopamine (6-OHDA) cause loss of catecholamines from neurones, and when injected into the third ventricle of rats there is a considerable rise in plasma corticosterone accompanying the loss of hypothalamic noradrenaline. Guanethedine given *systemically* has no such effects. It does not readily cross the blood–brain barrier. This provides further evidence that the important catecholamine fibres affecting CRF output do not end in the median eminence (Scapagnini *et al.*, 1972; Cuello *et al.*, 1974). However, there are some aminergic fibres terminating in the median eminence, since rats given 6-OHDA intraventricularly show marked electron microscopic degenerative changes in some axons of the external layer of the median eminence (Cuello *et al.*, 1974).

Stress might also be expected to alter hypothalamic content of noradrenaline, its precursors or the enzymes involved in its synthesis or metabolism. In fact acute immobilization stress reduces, within five minutes, hypothalamic catecholamine, yet there is a concomitant rise in ACTH and corticosterone. If the immobilization is continued for 150 minutes, amounts of catecholamines are still low, and by this time ACTH and corticosteroid have also dropped. Acute stress lasting for 20 minutes is associated with a significant reduction in noradrenaline content of the ventromedial nucleus and supra-optic nucleus. If sustained (150 minutes) immobilization occurs every day for 40 days, and the rats are then *immediately* killed, noradrenaline content of the supra-optic, dorso-medial, paraventricular nuclei, and of the median eminence, is greatly elevated, while with the treatment continuing for 39 days, and rats killed 24 hours later, levels of noradrenaline in these nuclei are similar to control animals. Presumably there is a very rapid readjustment of noradrenaline synthesis between successive daily bouts of stress (Kvetnansky *et al.*, 1976; see also Keim and Sigg, 1976). A drop has been observed in the noradrenaline content of the arcuate nucleus following acute stress of various sorts, and with repeated immobilization stress, the tyrosine hydroxylase within the arcuate nucleus increases. Since there are no noradrenergic nerve cell bodies in the arcuate nucleus, the changes in noradrenaline and tyrosine hydroxylase must be due to alteration in nerve terminals consequent upon changed neuronal activity proximal to the arcuate nucleus (Kobayashi *et al.*, 1976).

Whole brain dopamine-β-hydroxylase (DBH), the enzyme which catalyses the final step in the biosynthesis of noradrenaline, rises within 15 minutes of mild stress. The speed with which this occurs suggests that it is not due to new synthesis of DBH, but to activation of the enzyme from an inactive pool. Furthermore, while hypophysectomy diminished hypothalamic DBH, $ACTH_{1\text{-}24}$ and even (in hypophysectomized rats) $ACTH_{1\text{-}10}$ causes a rise in hypothalamic DBH and a fall (in the case of $ACTH_{1\text{-}24}$) in cerebellar DBH (van Loon, 1976). Perhaps such effects of ACTH on the hypothalamus might be realized in intact animals through short-loop feedback pathways (Szentagothai *et al.*, 1968).

A nagging problem in this consideration of noradrenaline and the release of CRF and ACTH exists however. Are metabolic processes which are transynaptically regulated by adrenergic mechanisms (and CRF release would come into this catagory) influenced by circulating catecholamines of adrenal medullary and neuronal origin, which are released in response to stress, or is there some mechanism which provides protection from such stress-induced increase in extracellular catecholamines? The answer seems to be that there is some such safeguard. One approach to this problem has employed the pineal gland, which contains N-acetyltransferase, important in the synthesis of melatonin, and which is regulated by release of noradrenaline from sympathetic nerve terminals ending on pineal parenchymal cells. The noradrenaline acts through β-receptors. Both stress, and removal of the superior cervical ganglia, cause levels of N-acetyl transferase in the pineal to rise. Stress in ganglionectomized rats causes an even greater increase in the enzyme. In such pineals there is nerve terminal degeneration. Is their absence (and the presence of noradrenaline in the plasma) the cause for the greater increase in N-acetyl transferase? If the preganglionic fibres to the superior cervical ganglia are severed (decentralization), the nerve terminals in the pineal will remain intact. In rats prepared in such a way, the increment in N-acetyl transferase following stress is less than in ganglionectomized animals, though nevertheless more than in intact rats: a decentralized pineal lacking sympathetic input, may become supersensitive, as are pineals from rats exposed to constant light (Klein and Parfitt, 1976). Thus the degree to which an organ receives a noradrenergic innervation, and has noradrenergic nerve endings, influences its stress response. A rich innervation will mean high uptake of noradrenaline from extracellular fluids, and its destruction to a large extent by mitochondrial monoamine oxidase.

At the level of the CRF neurone, extracellular noradrenaline may be taken up by the NA nerve endings there and inactivated. There is also the suggestion that noradrenergic endings have their own α-receptors (in addition to those on the post-synaptic side of the cleft). Noradrenaline from the extracellular fluid, accepted by these receptors, may dampen the release of noradrenaline from the same nerve endings (Langer, 1974).

Interaction between the CRF–ACTH–adrenocortical system, and the release of other trophic hormones of the anterior pituitary

The release of at least TSH and gonadotrophins from the anterior pituitary gland is thought to be influenced by the CRF–ACTH–adrenocortical system. There appears to be an inverse relationship between the secretion of these hormones and of ACTH, so that TSH and gonadotrophin secretion is inhibited by stress (Fortier *et al.*, 1970; Christian, 1971). The effect of stress upon gonadotrophin secretion, or other aspects of reproduction, is of more obvious immediate relevance in any consideration of population control.

Studies of wild populations of some mammals suggested that there are changes in longevity and fertility with altered population sizes (Chitty, 1952). It has been supposed that the periodic dramatic decline in population size of some species results from intra-specific strife, a powerful stressor having effects on reproduction (Christian, 1950, 1971; Chitty, 1952). In experimental populations of voles, alteration in several parameters of reproduction occur as population size changes, and certainly voles do attack each other under experimental conditions, the fighting being a sufficiently powerful stressor to cause marked adrenal enlargement and thymus involution, indicative of hyperadrenocorticism (Clarke, 1953, 1955, 1956). The possibility that adrenocortical hormones are involved in increased mortality at particular stages of population cycles is strongly suggested by study of the small dasyurid marsupial, *Antechinus stuartii* (Bradley *et al.*, 1976). All males in wild populations die within three weeks of the beginning of the August mating period. Males captured earlier and caged singly survive well beyond the time of natural mortality. Values for the total of free plus protein-bound corticosteroid are very much higher in August than July males, and than August females. The 'maximum corticosteroid binding capacity' (MCBC) of plasma protein of males dramatically declines in August, and the total plasma corticosteroid concentration greatly exceeds the MCBC. Females do not display such alterations in MCBC. It has been suggested that the widespread death of males following the mating period results from hyper-adrenocorticism, exacerbated by a fall in MCBC, changes caused by increased aggressiveness and interactions of the males.

On the other hand, observed changes in the size of adrenal glands of animals in natural or experimental populations cannot always be taken simply to indicate altered levels of stress. There is a well established, close, functional link between the adrenal cortex and the reproductive system, some parts of the cortex being influenced by gonadal hormones, and probably also by gonadotrophins and placental hormones. This applies to voles as well as to mice (Chester Jones, 1957; Chitty and Clarke, 1963; Jorné-Safriel, 1968). That the gonads or gonadal hormones may influence adrenocortical function has been suggested from studies on female rats which had been treated with testosterone at the age of five days. With this particular colony of rats plasma corticosterone levels were low at 08.00 and reached a peak at 18.00 h. Those treated with androgen on day 5 had higher levels of corticosterone at 08.00 and 18.00 than untreated rats, and while ovariectomized females treated with oestradiol have elevated corticosterone titres, ovariectomized, *androgenized* females do not show this response. Such experiments suggest that early treatment with testosterone alters adrenocortical sensitivity and increases cortocosteroid output, altering the prepubertal sequence of decline and rise in corticosterone levels, and abolishing the adrenocortical sensitivity to oestradiol. Furthermore, it has even been supposed that such exposure to androgen in early life (and the testosterone treatment could have its natural counterpart) may modify the response of the adrenal cortex to stress (Ramaley, 1976).

The exact manner in which stress affects reproduction is not clear. Sustained daily treatment of laboratory mice with supposed adrenal androgens or with ACTH, damages the ovaries, causing increased follicular atresia and interfering with ovulation. The reduced efficiency of reproduction associated with heightened social pressure and enhanced adrenocortical activity may be caused by inhibition of gonadotrophin release by adrenal androgens (Varon and Christian, 1963; Christian, 1964). Large doses of ACTH given to rats apparently interfere with PMS induced ovulation, suggesting that an adrenal product can exert a negative feedback action on the secretion of gonadotrophin (Hagino, 1968).

The clearest example of the effect of stress on reproduction is pregnancy block in mice ('Bruce' effect). If female mice are exposed to a strange male shortly after mating with a stud male, the pregnancy that might be expected from the initial mating fails. This is because implantation does not occur, probably as a result of deficient prolactin secretion (Bruce, 1967; Bruce and Parkes, 1960). Subjecting recently mated females to multiple short-term exposure to strange males for 3 × 15 minutes per day for four days causes the incidence of pregnancies to be much lower than in females not disturbed after mating (50% compared with 90%). Such short, repeated, exposure to strange males may approximate to conditions in natural populations (Chipman *et al.*, 1966). This pregnancy block can occur, under laboratory conditions, in voles and bank voles (Clulow and Clarke, 1968; Clarke and Clulow, 1973; Milligan, 1976a,b). It may be that some part of the decline in fertility of natural and experimental vole populations, correlated with high population density (Chitty, 1952; Clarke, 1955), is attributable to pregnancy block. Some facts relating to wild bank voles (*Clethrionomys glareolus*) and skomer voles (*Clethrionomys glareolus skomerensis*) are consistent with the idea that pregnancy block occurs in natural populations. Brambell and Rowlands (1936) and Coutts and Rowlands (1969) found a number of sets (or generations) of corpora lutea in these species, with as many as 17 corpora lutea in the most recent set. Ovulation in bank voles is induced by mating (Clarke *et al.*, 1970). Ovaries resembling those found in wild populations, with several generations of corpora lutea, some healthy, others already degenerating, have been produced in laboratory bank voles by exposing them to a succession of males, without allowing the females to survive to implantation. However, if implantation is permitted in such females, the numerous (up to 16) corpora lutea are all equally healthy histologically, although they vary considerably in size: implantation allows 're-capture' or corpora lutea induced by a first mating and switched off by a subsequent male (Clarke and Clulow, 1973).

That pregnancy block is at least partly due to activity of the CRF–ACTH–adrenocortical system has been clearly demonstrated in experiments on intact and adrenalectomized mice mated with a stud male, followed in some cases by exposure to a strange male. The occurrence of pregnancy block was greatly reduced in adrenalectomized mice. Unexpectedly, it was concluded that the mechanism of pregnancy block involved an effect of ACTH directly on the anterior pituitary, or indirectly through a higher centre, altering the release of gonadotrophin during critical pre-implantation stages of pregnancy (Snyder and Taggart, 1967). However the adrenalectomized animals, which presumably had high plasma titres of ACTH (see above, pp. 133–135) had a low incidence of pregnancy *failure*. It would seem more reasonable to suppose from these results that the adrenal cortex of female mice exposed to a strange male after mating with a stud male, releases materials which, acting on the uterus directly or indirectly through the ovaries or hypothalamo–hypophysial system, contribute to the pregnancy block. In addition, the deficiency of prolactin release may also contribute (Bruce, 1967), *via* stress activation of dopaminergic pathways ending in the medium eminence. It has been suggested that the prolactin-inhibiting factor (PIF) is dopamine (Ganong, 1974). Stress-induced increase of dopamine release into the primary capillary plexus in the median eminence could then interfere with the discharge of prolactin from the anterior pituitary.

The secretions of the adrenal medulla influence the outcome of pregnancy. Intact mice exposed daily throughout pregnancy to the emotional stress caused by the blockade of escape in a conditioned-avoidance situation involving electric shock, have a considerably higher incidence of still-births than mice treated in exactly the same way but which, prior to mating, have had their adrenal medullas removed (Caldwell, 1962).

It is clear that reproductive success is affected by stress. The physiological mechanisms by which activity of the CRF–ACTH–adrenocortical system (on the one hand), and that of the adrenal medulla (on the other), reduces fertility, need to be more fully investigated. The demographic significance of these processes is still largely conjectural, being extensively based on indirect evidence.

Does the stress response fulfil any useful function?

It is quite clear that in at least some of their effects, secretions of the adrenal gland increase the chances of survival of a stressed animal. The cardiovascular and metabolic effects of adrenal medullary secretions, and of their counterparts released from sympathetic nerve terminals, are of value to an animal whose life is threatened by changes in the internal or external environment. Similarly the effects of corticosteroids upon carbohydrate metabolism, and their interactions with adrenal medullary hormones in the regulation of blood pressure, are vital.

However other effects of corticosteroids are less easily interpreted, most notably their capacity to reduce inflammation, to reduce wound healing and granuloma formation, their reduction in the immune response and in the activity of macrophages. These effects have been observed in intact animals exposed to a variety of stressors, including emotional stress (Christian and Williamson, 1958; Kilbourne *et al.*, 1961; Johnson *et al.*, 1963; Funk and Jensen, 1967; Rasmussen, 1969; Solomon, 1969; Joasoo and McKenzie, 1976).

There does seem to be one supremely important role for the CRF–ACTH–corticosteroid system, at least in sheep and goats. It is now clear that parturition in sheep depends on the secretion of ACTH and thence corticosteroids (Liggins *et al.*, 1972; Thorburn *et al.*, 1972). Concomitantly levels of prostaglandin ($PGF_{2\alpha}$) rise sharply in maternal blood. It seems likely that $PGF_{2\alpha}$, acting on the myometrium, initiates the uterine contractions and parturition. It may be that this sequence of events is a response by the foetus to the stress of hypoglycaemia (Jones, 1976) or to the hot environment it has quite suddenly realized is its habitat (Thornburn *et al.*, 1972; G. Thornburn, personal communication). Until a week or two before birth, foetal lambs have an imperfectly developed thermoregulatory system, which may in part be due to inadequate thermoreceptors. For example, lambs born prematurely thermoregulate poorly. Foetal lambs, though having body temperatures 1–2 centigrade degrees above the mother, have thyroxine levels which are (unexpectedly) above those of the mother. In the last 7 to 10 days of foetal life, foetal thyroxine levels fall to values resembling the mother. Conceivably this is because of maturation of thermoreceptors. The detection by the foetus of its ambient temperature could amount to stress. The start of free, post-natal, life may be its most important effect.

References

Allen, J. P., Allen, C. F., Greer, M. A., and Jacobs, J. J. (1973) Stress induced secretion of ACTH. In *Brain-Pituitary-Adrenal Interrelationships*, ed. Brodish, A. and Redgate, E. S. pp. 99–127. Basel: Karger.

Azmita, C. E. and McEwen, B. S. (1974) Adrenalcortical influence on rat brain tryptophan hydroxylase activity. *Brain Research*, **78**, 291–302.

Bradley, A. J., McDonald, I. R., and Lee, A. K. (1976) Corticosteroid binding globulin and mortality in a dasyurid marsupial. *Journal of Endocrinology*, **70**, 323–324.

Brambell, F. W. R. and Rowlands, I. W. (1936) Reproduction of the bank vole (*Evotomys glareolus Schreber*). *Philosophical Transactions of the Royal Society, London, B*, **116**, 71–97.

Brodish, A. (1973) Hypothalamic and extra hypothalamic corticotrophin-releasing factors in peripheral blood. In *Brain-Pituitary-Adrenal Interrelationships*, ed. Brodish, A. and Redgate, E.S. pp. 128–151. Basel: Karger.

Bruce, H. M. and Parkes, A. S. (1960) Hormonal factors in exteroceptive block to pregnancy in mice. *Journal of Endocrinology*, **20**, xxix-xxx.

Bruce, H. M. (1967) Effects of olfactory stimuli on reproduction in mammals. In *Effects of External Stimuli on Reproduction*, ed. Wolstenholme, G. E. W. and O'Connor, M. pp. 29–42. London: Churchill.

Caldwell, D. F. (1962) Stillbirths from adrenal demedullated mice subjected to chronic stress throughout gestation. *Journal of Embryology and Experimental Morphology*, **10**, 471–475.

Callingham, B. A. (1975) Catecholamines in blood. In *Handbook of Physiology*, ed. Greep, R. O. and Astwood, E. B. Ch. 28; pp. 427–445. Washington: American Physiological Society.

Chester Jones, I. (1957) *The Adrenal Cortex*. London: Cambridge University Press.

Chipman, R. K., Holt, J. A., and Fox, K. A. (1966) Pregnancy failure in laboratory mice after multiple short term exposure to strange male. *Nature*, **210**, 653.

Chitty, D. H. (1952) Mortality among voles (*Microtus agrestis*) at Lake Vyrnwy, Montgomeryshire in 1936-9. *Philosophical Transactions of the Royal Society, London, B*, **236**, 505–552.

Chitty, H. and Clarke, J. R. (1963) The growth of the adrenal gland of laboratory and field voles and changes in it during pregnancy. *Canadian Journal of Zoology*, **41**, 1025–1034.

Christian, J. J. (1950) The adreno-pituitary system and population cycles in mammals. *Journal of Mammalogy*, **31**, 247–259.

Christian, J. J. (1964) Effect of chronic ACTH treatment on maturation of intact female mice. *Endocrinology*, **74**, 669–679.

Christian, J. J. (1971) Population density and reproductive efficiency. *Biology of Reproduction*, **4**, 248–294.

Christian, J. J. and Williamson, H. O. (1958) Effect of crowding on experimental granuloma formation in mice. *Proceedings of the Society of Experimental Biology and Medicine*, **99**, 385–387.

Clarke, J. R. (1953) The effect of fighting on the adrenals, thymus and spleen of the vole (*Microtus agrestis*). *Journal of Endocrinology*, **9**, 114–126.

Clarke, J. R. (1955) Influence of numbers on reproduction and survival in two experimental vole populations. *Proceedings of the Royal Society, London, B*, **144**, 68–85.

Clarke, J. R. (1956) The aggressive behaviour of the vole. *Behaviour*, **9**, 1–23.

Clarke, J. R. and Clulow, F. V. (1973) The effect of successive matings upon bank vole (*Clethrionomys glareolus*) and vole (*Microtus agrestis*) ovaries. In *The Development and Maturation of the Ovary and its Functions*, ed. Peters, H. pp. 160–170. Amsterdam: Excerpta Medica.

Clarke, J. R., Clulow, F. V., and Greig, F. (1970) Ovulation in the bank vole. *Journal of Reproduction and Fertility*, **23**, 531.

Clulow, F. V. and Clarke, J. R. (1968) Pregnancy block in *Microtus agrestis*, an induced ovulator. *Nature*, **219**, 511.

Coutts, R. R. and Rowlands, I. W. (1969) The reproductive cycle of the Skomer vole (*Clethrionomys glareolus skomerensis*). *Journal of Zoology*, **158**, 1–25.

Cuello, A. C., Shoemaker, W. J., and Ganong, W. F. (1974) Effect of 6-hydroxydopamine on hypothalamic nor-epinephrine and dopamine content, ultrastructure of the median eminence and plasma corticosterone. *Brain Research*, **78**, 57.

Dallman, M. F. and Jones, M. T. (1973a) Corticosteroid feedback control of stress induced ACTH secretion. In *Brain-Pituitary-Adrenal Interrelationships*, ed. Brodish, A. and Redgate, E. S. pp. 176–196, Basel: Karger.

Dallman, M. F. and Jones, M. T. (1973b) Corticosteroid feedback control of ACTH secretion: effect of stress-induced corticosterone secretion on subsequent stress response in the rat. *Endocrinology*, **92**, 1367–1375.

Dallman, M. F., Jones, M. T., Vernikos-Danellis, J., and Ganong, W. F. (1972) Corticosteroid feedback control of ACTH secretion: rapid effects of bilateral adrenalectomy on plasma ACTH in the rat. *Endocrinology*, **91**, 961–968.

Feldman, S. (1973) The interaction of neural and endocrine factors regulating hypothalamic activity. In *Brain-Pituitary-Adrenal Interrelationships*, ed. Brodish, A. and Redgate, E. S. pp. 224–238. Basel: Karger.

Feldman, S., Conforti, N., Chowers, I., and Davidson, J. M. (1970) Pituitary-adrenal activation in rats with medial basal hypothalamic islands. *Acta Endocrinologica, Copenhagen*, **63**, 405–414.

Feldman, S., Conforti, N., and Chowers, I. (1971) The role of the medial forebrain bundle in mediating adrenocortical response to neurogenic stimuli. *Journal of Endocrinology*, **51**, 745.

Feldman, S., Conforti, N., and Chowers, I. (1972) Neural pathways mediating adrenocortical responses to photic and acoustic stimuli. *Neuroendocrinology*, **10**, 316–323.

Fortier, C., Delgado, A., Ducommun, P., Ducommun, S., Dupont, A., Jobin, M., Kraicer, J., MacIntosh-Hardt, B., Marceau, H., Mialhe, P., Mialhe-Voloss, C., Rerup, C., and van Rees, G. P. (1970) Functional interrelationships between the adenohypophysis, thyroid, adrenal cortex and gonads. *Canadian Medical Association Journal*, **103**, 864–874.

Funk, G. A. and Jensen, M. M. (1967) Influence of stress on granuloma formation. *Proceedings of the Society for Experimental Biology and Medicine*, **124**, 653–655.

Fuxe, K. and Hökfelt, T. (1969) Catecholamines in the hypothalamus and the pituitary gland. In *Frontiers in Neuroendocrinology, 1969*, ed. Ganong, W. F. and Martini, L., pp. 47–96. New York: Oxford University Press.

Fuxe, K., Hökfelt, T., Jonsson, G., Levine, S., Lidbrink, P., and Löfström, A. (1973) Brain and pituitary-adrenal interactions. Studies on central monoamine neurones. In *Brain-Pituitary-Adrenal Interrelationships*, ed. Brodish, A. and Redgate, E. S. pp. 239–269. Basel: Karger.

Gann, D. S. and Cryer, G. L. (1973) Feedback control of ACTH secretion by cortisol. In *Brain-Pituitary-Adrenal Interrelationships*, ed. Brodish, A. and Redgate, E. S. pp. 197–223, Basel: Karger.

Ganong, W. F. (1963) The central nervous system and the synthesis and release of adrenocorticotropic hormone. In *Advances in Neuroendocrinology*, ed. Nalbandov, A.V. pp. 92–157. Urbana: University of Illinois Press.

Ganong, W. F. (1974) The role of catecholamines and acetylcholine in the regulation of endocrine function. *Life Science*, **15**, 1404–1414.

Ganong, W. F., Wise, B. L., Shackelford, R., Boryczka, A. T., and Zipf, B. (1965) Site at which α-ethyltryptamine acts to inhibit the secretion of ACTH. *Endocrinology*, **76**, 526–530.

Ganong, W. F., Kramer, N., Reid, I. A., Boryczka, A. T., and Shackelford, R. (1976) Inhibition of stress-induced ACTH secretion by norepinephrine in the dog: mechanism and site of action. In *Catecholamines and Stress. Proceedings of the International Symposium on Catecholamines and Stress, Bratislava*. pp. 139–143. Oxford: Pergamon Press.

Gibbs, F. P. (1969) Area of pons necessary for traumatic stress-induced ACTH release under pentobarbital anaesthesia. *American Journal of Physiology*, **217**, 84–87.

Gordon, R., Spector, S., Sjoerdsma, A., and Udenfriend, S. (1966) Increased synthesis of norepinephrine and epinephrine in the intact rat during exercise and exposure to cold. *Journal of Pharmacology and Experimental Therapeutics*, **153**, 440–447.

Grosser, B. I., Stevens, W., and Reed, D. J. (1973) Properties of corticosterone-binding macromolecules from rat brain cytosol. *Brain Research*, **57**, 387–395.

Hagino, N. (1968) Inhibition of gonadotrophin-induced ovulation by ACTH in immature female rats. *Excerpta Medica International Congress Series*, **157**, 61.

Harris, G. W. (1955) *Neural Control of the Pituitary Gland*. London: Edward Arnold.

Harris, G. W. (1972) Humours and hormones. *Journal of Endocrinology*, **53**, ii-xxii.

Haus, E. and Halberg, F. (1970) Circannual rhythm in level and timing of serum corticosterone in standardized inbred mature C-mice. *Environmental Research*, **3**, 81–106.

Hillhouse, E. W. and Jones, M. T. (1976) Effect of bilateral adrenalectomy and corticosteroid therapy on the secretion of corticotrophin-releasing factor activity from the hypothalamus of the rat *in vitro*. *Journal of Endocrinology*, **71**, 21–30.

Hökfelt, T. and Fuxe, K. (1972) On the morphology and the neuroendocrine role of the hypothalamic catecholamine neurons. In *Brain-Endocrine Interaction. Median eminence: structure and function*, ed. Knigge, K. M., Scott, D. E., and Weindl, A. pp. 181–223. Basel: Karger.

Huttunen, M. O. (1971) Persistent alteration of turnover of brain noradrenaline in the offspring of rats subjected to stress during pregnancy. *Nature*, **130**, 53–55.

Joasoo, A. and McKenzie, J. M. (1976) Stress and the immune response in rats. *International Archives of Allergy and Applied Immunology*, **50**, 659–663.

Johnson, T., Lavender, J. F., Hultin, E., and Rasmussen, A. F. (1963) The influence of avoidance-learning stress on resistance to Cocksackie B virus in mice. *Journal of Immunology*, **91**, 569–575.

Jones, C.T. (1976) Hypoglycaemia as a stimulus for adrenocorticotrophin secretion in foetal sheep. *Journal of Endocrinology*, **70**, 321–322.

Jorné-Safriel, O. (1968) Some factors affecting the adrenal juxtamedullary zone in the vole (*Microtus agrestis*) and bank vole (*Clethrionomys glareolus*). D.Phil. thesis, University of Oxford.

Keim, K. L. and Sigg, E. B. (1976) Physiological and biochemical concomitants of restraint stress in rats. *Pharmacology Biochemistry and Behaviour*, **4**, 289–297.

Kendall, J. W. (1971) Feedback control of adrenocorticotrophic hormone secretion. In *Frontiers in Neuroendocrinology*, ed. Martini, L. and Ganong, W. F. pp. 177–207. New York: Oxford University Press.

Kilbourne, E. D., Smart, K. M., and Pokorny, B. A. (1961). Inhibition by cortisone of the synthesis and action of interferon. *Nature*, **190**, 650–651.

Klein, D. C. and Parfitt, A. (1976) A protective role of nerve endings in the stress-stimulated increase in pineal N-acetyltransferase activity. In *Catecholamines and Stress. Proceedings of the International Symposium on Catecholamines and Stress, Bratislava*. pp. 119–128. Oxford: Pergamon Press.

Kobayashi, R. M., Palkovits, M., Kizer, J. S., Jacobowitz, D. M., and Kopin, I. J. (1976) Selective alterations of catecholamines and tyrosine hydroxylase activity on the hypothalamus following acute and chronic stress. In *Cate-*

cholamines and Stress. Proceedings of the International Symposium on Catecholamines and Stress, Bratislava. pp. 29–38. Oxford: Pergamon Press.

Kvetnansky, R., Mitro, A., Palkovits, M., Brownstein, M., Torda, T., Vigas, M., and Mikulaj, L. (1976) Catecholamines in individual hypothalamic nuclei in stressed rats. In *Catecholamines and Stress. Proceedings of the International Symposium on Catecholamines and Stress, Bratislava*. pp. 39–50. Oxford: Pergamon Press.

Langer, S. Z. (1974) Presynaptic regulation of catecholamine release. *Biochemical Pharmacology*, **23**, 1793–1800.

Lewis, G. P. (1975) Physiological mechanisms controlling secretory activity of adrenal medulla. In *Handbook of Physiology*, ed. Greep, R. O. and Astwood, E. B. Ch. 22, pp. 309–319. Washington: American Physiological Society.

Liggins, G. C., Grieves, S. A., Kendal, J. Z., and Knox, B. S. (1972) The **physiological roles of progesterone, oestradiol—17β and prostaglandin $F_{2\alpha}$** in the control of ovine parturition. *Journal of Reproduction and Fertility Supplement*, **16**, 85–103.

van Loon, G. R. (1976) Brain dopamine betahydroxylase activity: response to stress, tyrosine hydroxylase inhibition, hypophysectomy and ACTH administration. In *Catecholamines and Stress. Proceedings of the International Symposium on Catecholamines and Stress, Bratislava*. pp. 77–87. Oxford: Pergamon Press.

Matheson, G. K., Branch, B. J., and Taylor, A. N. (1971) Effect of amygdaloid stimulation on pituitary-adrenal activity in conscious cats. *Brain Research*, **32**, 151–167.

McEwan, B. S. and Wallach, S. (1973) Corticosterone binding to hippocampus: nuclear and cytosol binding *in vitro*. *Brain Research*, **57**, 373–386.

McEwan, B. S., Weiss, J. M., and Schwartz, L. S. (1969) Uptake of corticosterone by rat brain and its concentration by certain limbic structures. *Brain Research*, **16**, 227–241.

McEwan, B. S., Denef, C. J., Gerlach, J. L., and Plapinger, L. (1974) Chemical studies on the brain as a steroid hormone target tissue. In *The Neurosciences. Third Study Program*, ed. Schmitt, F. O. and Worden, F. G. pp. 599–620. Cambridge, Mass.: MIT Press.

McNatty, K. P., Cashmore, M., and Young, A. (1972) Diurnal variation in plasma cortisol levels in sheep. *Journal of Endocrinology*, **54**, 361–362.

Michel, E. K. (1974) Dexamethasone inhibits multiunit activity in the rat hippocampus. *Brain Research*, **65**, 180–183.

Milligan, S. R. (1976a) Pregnancy blocking in the vole, *Microtus agrestis*. I. Effect of the social environment. *Journal of Reproduction and Fertility*, **46**, 91–95.

Milligan, S. R. (1976b) Pregnancy blocking in the vole, *Microtus agrestis*. II. Ovarian, uterine and vaginal changes. *Journal of Reproduction and Fertility*, **46**, 97–100.

Palkovits, M. and Stark, E. (1972) Quantitative histological changes in the rat hypothalamus following bilateral adrenalectomy. *Neuroendocrinology*, **10**, 23–30.

Palkovits, M., Brownstein, M., Kizer, J. S., Saavedra, J. M., and Kopin, I. J. (1976) Effect of stress on serotonin and tryptophan hydroxylase activity of brain nuclei. In *Catecholamines and Stress. Proceedings of the International Symposium on Catecholamines and Stress, Bratislava*. pp. 51–59. Oxford: Pergamon Press.

Ramaley, J. A. (1976) Serum corticosterone in rats with delayed anovulation. *Journal of Endocrinology*, **71**, 31–36.

Rasmussen, A. F. (1969) Emotions and immunity. *Annals of the New York Academy of Science*, **164**, 458–462.

Redgate, E. S., Fahringer, E. E., and Szechtman, H. (1973) Effects of the nervous system on pituitary adrenal activity. In *Brain-Pituitary-Adrenal Interrelationships*, ed. Brodish, A. and Redgate, E. S. pp. 152–175. Basel: Karger.

Rees, L. H., Cook, D. M., Kendall, J. W., Allen, C. F., Kramer, R. M., Ratcliffe, J. G., and Knight, R. A. (1971) A radioimmunoassay for rat plasma ACTH. *Endocrinology*, **89**, 254–261.

Sato, T., Sato, M., Shinako, J., and Dallman, M. F. (1975) Corticosterone induced changes in hypothalamic corticotropin-releasing factor (CRF) content after stress. *Endocrinology*, **97**, 265–274.

Scapagnini, U., van Loon, G. R., Moberg, G. P., Preziosi, P., and Ganong, W. F. (1972) Evidence for central norepinephrine mediating inhibition of ACTH secretion in the rat. *Neuroendocrinology*, **10**, 155–160.

Segal, M. (1976) Interactions of ACTH and norepinephrine on the activity of rat hippocampal cells. *Neuropharmacology*, **15**, 329–333.

Segal, M. and Bloom, F. E. (1974a) The action of norepinephrine in the rat hippocampus. I. Iontophoretic studies. *Brain Research*, **72**, 79–97.

Segal, M. and Bloom, F. E. (1974b) The action of norepinephrine in the rat hippocampus. II. Activation of the input pathway. *Brain Research*, **72**, 99–114.

Sharp, G. W. G., Sloroch, S. A., and Vipond, H. J. (1961) Diurnal rhythms of the keto- and ketogenic steroid excretion and the adaptation to changes of the activity-sleep routine. *Journal of Endocrinology*, **22**, 377–385.

Snyder, R. L. and Taggart, N. E. (1967) Effects of adrenalectomy on male induced pregnancy block in mice. *Journal of Reproduction and Fertility*, **14**, 451–455.

Solomon, G. F. (1969) Stress and antibody response in rats. *International Archives of Allergy and Immunology*, **35**, 97–104.

Stevens, W., Grosser, B. I., and Reed, D. J. (1971) Corticosterone-binding molecules in rat brain cytosols: regional distribution. *Brain Research*, **35**, 602–607.

Stone, E. A. (1973) Adrenergic activity in rat hypothalamus following extreme muscular exertion. *American Journal of Physiology*, **224**, 165–169.

Szentagothai, J., Flerko, B., Mess, B., and Halasz, B. (1968) *Hypothalamic Control of the Anterior Pituitary. An experimental-morphological study*. 3rd edition. Budapest: Akademiai Kiado.

Telegdy, G. and Vermes, I. (1976) Changes induced by stress in the activity of the serotoninergic system in the limbic brain structures. In *Catecholamines and Stress. Proceedings of the International Symposium on Catecholamines and Stress, Bratislava*. pp. 145–156. Oxford: Pergamon Press.

Thierry, A. W., Javoy, F., Glowinski, J., and Kety, S. S. (1968) Effects of stress on the metabolism of norepinephrine, dopamine and serotonin in the central nervous system of the rat. I. Modifications of norepinephrine turnover. *Journal of Pharmacology and Experimental Therapeutics*, **163**, 163–171.

Thierry, A. M., Blanc, G., and Glowinski, J. (1970) Preferential utilization of newly synthesised norepinephrine in the brain stem of stressed rats. *European Journal of Pharmacology*, **10**, 139–142.

Thoenen, H. (1970) Induction of tyrosine hydroxylase in peripheral and central adrenergic neurones by cold exposure of rats. *Nature (London)*, **228**, 861–862.

Thornburn, G. D., Nicol, D. H., Bassett, J. M., Shutt, D. A., and Cox, R. I.

(1972) Parturition in the goat and sheep: changes in corticosteroids, progesterone, oestrogens and prostaglandin F. *Journal of Reproduction and Fertility Supplement*, **16**, 61–84.

Tiptaft, E. and Jones, M. T. (1976) Effects of 18-hydroxydeoxy-corticosterone on stress-induced release of corticotrophin. *Journal of Endocrinology*, **69**, 33P–34P.

Ungerstedt, U. (1974) Functional dynamics of central monoamine pathways. In *The Neurosciences. Third Study Programme*, ed. Schmitt, F. O. and Worden, F. G. pp. 979–988. Cambridge, Mass.: MIT Press.

Varon, H. H. and Christian, J. J. (1963) Effects of adrenal androgens on immature female mice. *Endocrinology*, **72**, 210–222.

Yates, F. E. and Maran, J. W. (1974) Endocrinology. In *Handbook of Physiology, Vol. IV The pituitary gland and its neuroendocrine control*. ed. Sawyer, W. and Knobil, E. Section 7, Part 2. Washington: American Physiological Society.

Yates, F. E., Russell, S. M., and Maran, J. W. (1971) Brain-adrenohypophysial communication in mammals. *Annual Review of Physiology*, **33**, 393–444.

The experimental analysis of overcrowding

J. H. MACKINTOSH

Sub-Department of Ethology, University of Birmingham Medical School, Birmingham, England

Introduction

Reasons for the widespread interest in the phenomena associated with overcrowding are not hard to find. We have become only too aware of the rate of increase of human populations and often we feel that already there are just too many people about. The experience of crowding is considered unpleasant, at least in Western society and we are apprehensive of its effects, particularly as we know that the situation will inevitably get worse. Even without this sort of motivation the study of the regulation of animal numbers is clearly of great importance and there has therefore been a search for a laboratory model of overcrowding. Usually the animal selected for this purpose has been a rodent, e.g. by Calhoun (1961, 1962), the choice being, to some extent, that of Hobson. Only a mammal will provide a plausible base for those interested in extrapolation and it has to be small in order to fit crowded populations into the space that most laboratories have available. Rodents are also attractive for this purpose because it is well known that in some species, in some habitats, the population periodically reaches very high densities and thereafter crashes (Elton, 1942; Chitty, 1952; etc.). The model therefore might show, not only the characteristics of a population which is increasing rapidly, but also those features which terminate its growth and bring about its downfall.

Experiments with dense populations of rodents have shown many interesting physiological concomitants and these have been extensively reviewed on previous occasions (e.g. Christian, 1963; Archer, 1970; Brain, 1975). I am therefore not going to refer to them in detail in this paper; rather I intend to consider some of the problems that exist in establishing a rodent model of overcrowding and to look in detail at the way in which available space affects the behaviour of one rodent, the house mouse.

Overcrowding is not easily defined. Presumably it is one stage worse than 'crowding' defined by Brain (1975) as 'a condition in which the individuals are subjectively judged by the observer to be in a condition well beyond their natural housing density'; it is in addition a value judgement carrying implications that what is happening to the population is a 'bad thing'. It is preferable to consider, rather, the effects of high densities and of limitations in available space. This has a number of advantages. Firstly, it emphasizes that the situation being examined is part of a continuum: if there are high densities there are also by inference low ones. Secondly, it draws attention to two of the components of crowding, namely the large numbers of individuals and the small amounts of living space, two variables which are by no means necessarily equivalent. Experiments in this area fall into two broad categories. Either a breeding colony is allowed to expand with unlimited resources, usually in a substantial enclosure, or a crowded group is set up directly, often in a normal sized cage. The former is clearly more interesting as it is more similar to a natural situation and here we are talking not only of overcrowding, but of the establishment of an experimental population.

Natural and experimental populations

It seems to be a reasonable proposition that an experimental population is only useful if it reflects, initially at least, what is known of the population structure and social behaviour of the species under natural conditions. With many rodents an assessment of the degree of concordance between two situations may itself present difficulties.

There have been many studies of populations in the field but numerical results are not necessarily exact. It is generally impossible to count all the animals at once, so that sampling techniques have to be used and these are not all noted for their precision or reliability. The tendency for capture–recapture techniques to reflect trapability, rather than actual numbers, is only too well known and biases produced by initial responses to traps may become aggravated or even reversed by repeated sampling. Some interference with the population is implicit in most methods even if it is no more than the frequent presence of the investigator, e.g. traps provide cover that was previously absent and some degree of provisioning is often unavoidable. The result of this is that, even if the technique itself is reliable, the population measured may differ significantly from that which was there before the start of operations. The sampling process may also be disturbed by changes in condition, temperature, humidity, etc., or by outside interference so that again an inaccurate estimate of numbers is made.

In addition, under field conditions the values of the various parameters affecting the population are seldom known with any precision. Resources and factors leading to losses from the population all have to be estimated.

A more serious problem is that behavioural field studies are very infrequent, particularly of those species which are most used in population investigations. This is not surprising as most rodents in this category are small, nocturnal, and cryptic in habit, so that direct observation is extremely difficult. Often the only way, therefore, of checking that the social organization seen in the laboratory is not an artefact of the experimental conditions is to measure it against deductions from trapping results.

Many of the difficulties inherent in field studies are absent in the laboratory. The amounts of food, water, and cover available are known, temperature and humidity can be controlled within narrow limits, and interference other than that inherent in the experiment can be excluded. Similarly, there is no doubt about the size, age structure, and sex ratio of the starting population; with a confined space intermediate estimates of numbers are not difficult and a final count presents no problems.

In spite of all these advantages, the laboratory situation presents of course a major problem. The environment is likely to be very different from the natural one and this discrepancy becomes worse with increasing ecological and behavioural sophistication. Even if every attempt is made to construct a representation of the natural conditions, there are found to be many differences. Worse still, some of the differences that exist may be unrecognized and certainly the effects, even of those which are known to be present, may be obscure.

Three of the major variables involved are environmental

complexity, available space, and social history of the founding stock.

Environmental complexity

It has been recognized on a number of occasions that a major feature of the experimental situation is its simplicity (e.g. Crowcroft, 1966). This characteristic is likely to interact with spatial variables and Anderson and Hill (1965), for example, suggested that a complex environment in necessary for territory formation in mice. Our results indicate that, within limits, this is not so. In trials with completely featureless enclosures no territories were formed, but the introduction of a few objects such as retort stands and wire boxes, or the provision of an opportunity to burrow, resulted in the establishment of defended areas.

A related pitfall in the simulation of a natural environment is that the structure that is produced is based on what the experimenter thinks is the natural situation, and he can be wrong. The knowledge that mice live in holes, for example, can lead to the assumption that the experimental situation should be an interconnecting system of tubes. Scott (1944) used an arrangement of this sort and failed to demonstrate territory formation in house mice, whereas field evidence suggests that these animals are normally territorial. Experimenters who have used a more open situation (e.g. Crowcroft, 1955a, 1955b; Crowcroft and Rowe, 1963; Mackintosh, 1970, 1973) have had no difficulty in producing territorial behaviour. The difference between Scott's results and the others may of course be due to other factors. Given an opportunity, e.g. an enclosure equipped with a false floor, similar to that used by van Oortmerssen (1970) (figure 1), mice will make a complex system of tunnels as would be expected, and these are incorporated into the territorial system, which, however, in our situation always extended into open areas beyond them. It may be that there is a difference between holes mice make themselves and those which are imposed upon them, or more probably it is the idea that a mouse's environment consists entirely of holes that is wrong. It must be pointed out, however, that Reimer and Petras (1967) used a multiple escape pen largely consisting of tunnels and reported territorial behaviour.

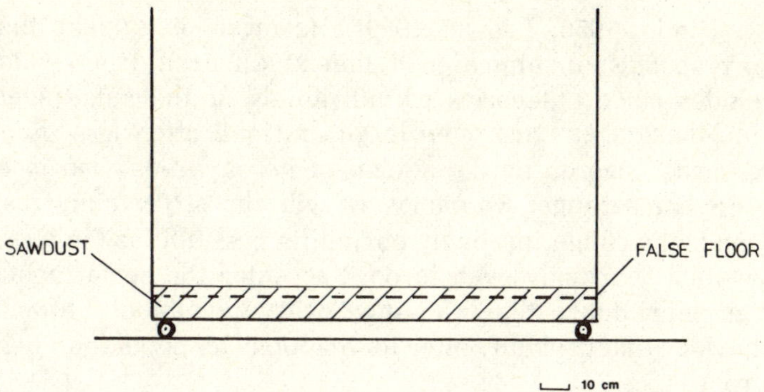

Figure 1 Section through enclosure with false floor.

The effects of limited space

Perhaps the most important feature or failing of the restricted environment is that it *is* restricted. The area involved is almost invariably small and emigration and immigration are impossible. Apart from island populations, which compared with those of the laboratory have a large area available, and a few special cases where small habitable regions are separated by large zones of very adverse environment, this is not true of natural populations.

The ways in which animals structure the space that surrounds them are complex and vary from species to species. Hediger (1950) drew attention to the major difference between what he called 'contact' species, which allow close physical proximity between animals apart from that involved in sexual, aggressive, or maternal behaviour, and 'distance' species which do not. He developed the concept of individual distance, an area around an animal within which it normally does not tolerate conspecifics, and Chance (1956) used a similar idea, in this case related to social structure, in describing the zone surrounding dominant rhesus monkeys. Later work has extended and illustrated the complexity of these relationships (e.g. Crook, 1961; McBride, 1971).

Perhaps the most spectacular spatial behaviour is territorial defence. Again, the behaviour covered by the term is diverse and numerous reviews have defined and redefined the concept (e.g. Hinde, 1956; Brown, 1969). However, it is clear that in many species an animal will defend an area from which it excludes some categories of individuals of the same species. Often these areas are quite large, but still often less than the total area covered by the animal in its day-to-day movements, i.e. its home range. An enclosure will almost certainly restrict an animal's range, probably curtail its possibilities for territory formation, and may even force it to enter the social space of other individuals. Each change will undoubtedly affect its behaviour and probably alter its reactions as population density increases.

The effects of previous experiences

Another problem facing the establishment of a laboratory population is that of the history of the founding stock. If caught wild, they have inevitably been through a process which under most definitions must be regarded as stressful and which may therefore affect later behaviour. Similarly their previous experiences, which in this situation are unlikely to have been recorded, will affect their behaviour under the experimental conditions. Effects of early experience on subsequent behaviour are too well known to be detailed, but a relevant example is provided by Bergerud and Hemus (1975) who showed that the behaviour of blue grouse in an experimental island population varied with the population density of the parent population from which the founding stock was drawn.

Laboratory breeding may produce more consistent subjects, but again the system used will affect the final outcome. Figure 2 illustrates an effect of different rearing conditions on 'Distance Ambivalence' in mice.

'Distance Ambivalence' is a category of behaviour, which includes a number of elements which are clearly vectors between approach and retreat (Grant and Mackintosh, 1963). In the mouse it shows strong sequential links with aggressive behaviour but with no other category and its frequency is therefore given in the figure relative to the amount of aggression shown. Those mice which were brought up and tested in laboratory cages (Group I) showed much less 'Distance Ambivalence' than was recorded in encounters in an enclosure (Group III). The difference, however, was not entirely an effect of the

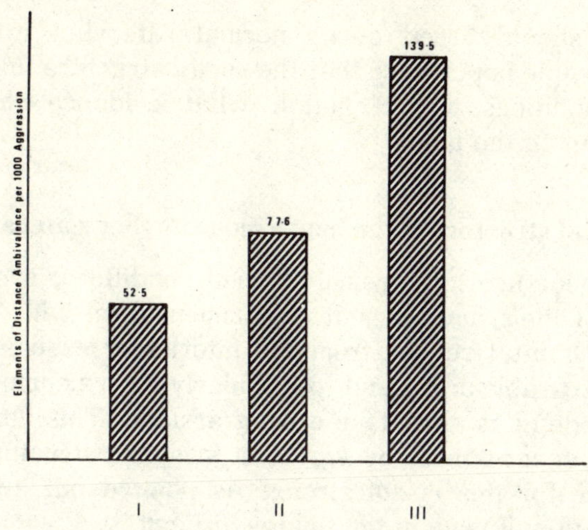

Figure 2 Frequency of elements of 'Distance Ambivalence' in mice: I bred and tested in cages; II bred in enclosures and tested in cages; III bred and tested in enclosures. N = 24 groups I and II, 15 group III.

test situation as the Group II mice, which had been raised in the enclosure but transferred to cages several weeks before testing, still showed a significantly high level of this behaviour compared with Group I.

Another example from our studies of the change of behaviour produced by caging is provided by the lack of a 'home cage effect' on agonistic behaviour, in a series of 120 fifteen-minute paired encounters. In each of these a mouse was introduced into the home cage of another with which it was unfamiliar and it was found that the home animal had no significant advantage over the intruder in terms of its score of aggressive behaviour. This is in marked contrast to the enclosure situation where the territorial mouse inevitably has the advantage and even mice which do not attain territorial status frequently show site-attached aggression and at this site they are dominant to many other individuals.

It follows, therefore, that unless both the behaviour of the animal under its natural conditions and its performance in the laboratory are reasonably well known, it is difficult to assess the significance of the laboratory experiment. The minimum criterion for the success of a laboratory population is that the animal should breed at its normal rate, but it is also of considerable importance that the social structure under laboratory conditions should match what evidence there is for behaviour in the field.

The social structure of the house mouse under natural conditions

House mice live in extremely varied conditions, probably as a result of their development of commensalism with man. Their normal habitat ranges from the interior of carcasses in cold stores to arable crops and, particularly where competition from other rodents is absent, to open grassland. This flexibility may well be accompanied by variation in social structure although evidence for this is not strong. As pointed out above, direct observation of mice in the field is difficult.

Extensive trapping studies have been carried out on field populations and these give some hint as to the probable social organization. Berry (1968, 1970), for example, has shown that some males show marked site attachment and these tend to be the large animals, younger males being more mobile. The static males hold their positions for long periods and if one of them disappears, then it is replaced, presumably by one of the itinerants.

The assumption, then, is that the site-attached males are territorial, although direct evidence is extremely sparse. A single field observation is illustrated in figure 3. Previous trapping results had indicated the presence of site-attached males at points A, B, and C, and a typical territorial encounter was seen at point X. This was of particular interest as laboratory results (Mackintosh, 1973) showed that landmarks are important in determining the site of the boundary between two territories and at X there was a conspicuous rock projecting from the wall.

House mouse behaviour in laboratory enclosures

As mentioned above, territory formation occurs readily as long as mice are given a large enough space. However, even if the area is sufficient for territory formation to take place the characteristics of the territories still vary to some extent with

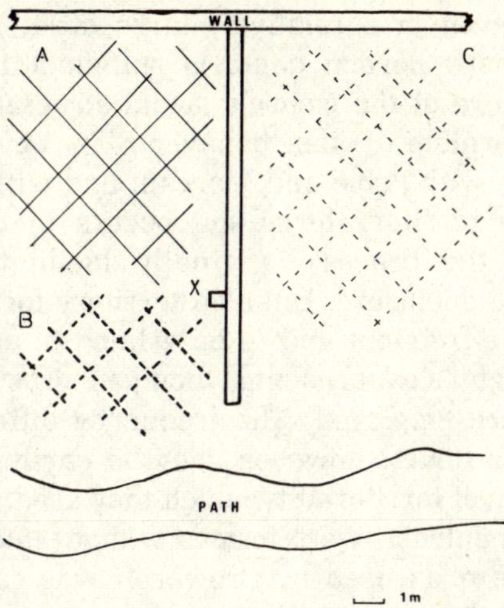

Figure 3 Diagram of field observation of territorial behaviour. Resident mice were located in areas A, B, and C. Encounter observed at point X.

spatial factors. Crowcroft, who was first to demonstrate territory formation in the mouse, used a large area for some of his experiments, and in these he showed a situation in which territories were held by individual males and these were partially separated by areas of no-man's land (Crowcroft, 1955b). The neutral zones contained many of the resources, such as food and watering points, and the territory holders met there apparently without conflict. The smaller enclosures that we have used and the similar sized ones used by Poole and Morgan (1976) produce a somewhat different picture as the territories took up all the individual space and therefore each individual territory must have contained adequate resources.

The difference in available space is also probably the cause of another difference between Crowcroft's findings and our own, in that, in his results, one or two females tend to be more or less permanently associated with each territorial male whereas in our case females are free to wander over the whole enclosure. Crowcroft's results match the field data best, e.g. resident females and young would produce a similar picture to that shown by Eibl-Eibesfeldt (1950) who stated that mice exist within territories as Grosfamilien. So that it is quite clear that confinement, even in a relatively large area, more than fifty times as big as a normal cage, is sufficient to eradicate an important feature of the mouse's social structure. Yet another trend is discernible if the behaviour in our enclosures is compared both with Poole and Morgan, and with Crowcroft.

Spontaneous territory formation occurs in our enclosures, (3.2 m^2), but the process is greatly facilitated by initially partitioning the enclosure. Unaided territory formation appears to become more frequent and to be related to enclosure size in Poole and Morgan's experiments when pen sizes of 1.3, 2.6, 3.8, and 5.2 m^2 were employed. The frequency difference between their results and ours, however, may be partly a result of the wider definition of territoriality which they used. All Crowcroft's territories, in contrast, were formed without intervention.

Although the area used by Crowcroft was sufficiently large for females to behave normally, his non-territorial males huddled together in a single nest box. Individuals in this category in the field are trapped on successive occasions at widely separate points and apparently do not congregate together.

Social differentiation of non-territorial males

The standard method that we have used to provoke territory formation is to place four adult male mice on each side of a partitioned enclosure 1.8 m × 1.8 m, for one week before the partition is removed. This usually leads to the establishment of two territories and the surplus males are subordinate to the territorial mice. There is, however, a degree of differentiation amongst these subordinate individuals. Some are aggressive towards other subordinates and form a sufficiently distinct category for them to be distinguished as subdominants. This group remains anatomically and probably physiologically unlike territory holders and similar to subordinates (Evans and Mackintosh, 1976).

Territorial males are characterized by the possession of large preputial glands, higher body weight, and, when assessed at one week after the establishment of territories, smaller adrenal glands.

Further observation has shown (figure 4) that the adrenal weight difference disappears after one month, indicating adapta-

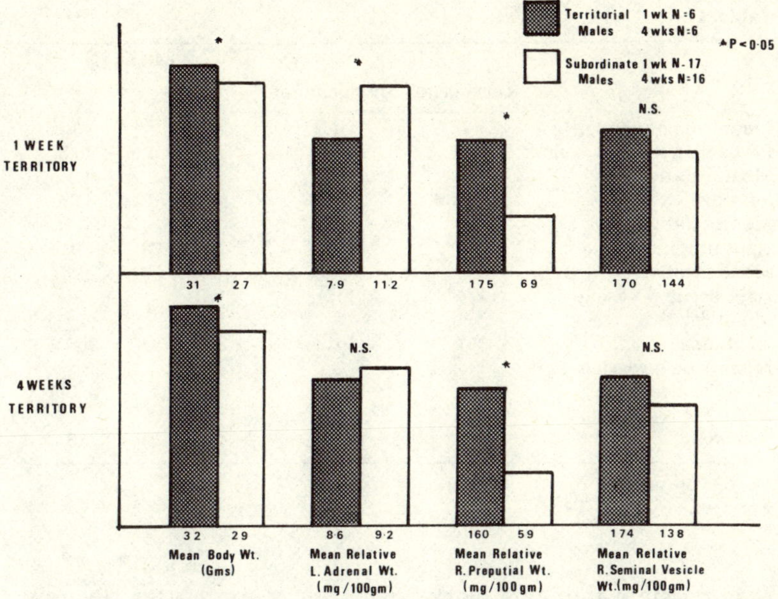

Figure 4 Organ weight changes in territorial and subordinate mice one week and four weeks after establishment of territories.

tion on the part of the subordinates; the preputial weight difference, however, is maintained.

In table 1 the remaining subordinate group is subdivided on the basis of coat condition into two groups, 'A' with little or no bite damage and 'B' with appreciable scarring, a distinction that is easy to make. The A group have significantly larger seminal vesicles than either the subdominants or the B group and there are indications that, although undoubtedly subordinate, they resemble the dominant group in some respects, e.g. in the proportion losing weight and possibly in adrenal weight. It is also clear that scarring and aggression received are not correlated.

The frequency of occurrence of the subdominant group varies, but there are usually at least one and often two in each trial.

The characteristics which define the four behavioural types found in the territorial situation are now being investigated, and

Table 1

	Territorial	Subdominant	Subordinate A	Subordinate B
Mean weight change/g	+0.7	-1.2	-0.4	-3.2
Proportion losing weight	1/6	5/6	1/5	5/5
Mean relative left adrenal	8.6	9.3	8.5	9.8
Mean relative right preputial	166**	61	59	58
Mean relative right seminal vesicle	174*	137	170	109
Mean relative right testis	251	261	300	260
Mean % aggression given/mouse	40	8	1.1	0.4
Mean % aggression received	<1	17	25	13

** P = < 0.01
* P = < 0.02 ANOVA

it is possible that this work will show even greater diversity. There are indications, for example, that the subdominant group itself may be subdivided on the basis of their relationships to the territory-holding mice. Some are subject to a considerable amount of aggression from the dominant; in fact they may be attacked significantly more often than some subordinates. Others are attacked very infrequently and these may even support the territory holder in defending his territory.

Subdivision of subordinate individuals into several categories has also been reported in mice by Henry *et al.* (1975) who detected three types of mouse in multiple escape pens, 'dominants', 'rivals', and 'subordinates'. Clarke (1955) showed a similar situation in colonies of *Microtus agrestis* where there were three types of male, some heavy and sleek—the dominants, others scarred and unkempt, and a third group intermediate between the other two. Barnett (1963) also distinguished between Beta and Omega classes of subordinate.

Differentiation of a population into different behavioural types has been described by Armitage (1975) in marmots and he suggested that the growth of populations in this species is affected by the frequency of the different behavioural categories present.

Dispersal of mice from laboratory populations

There is probably no general solution to the restriction on dispersion implicit in laboratory experiments, but a chance observation led us to a series of experiments in which we were at least partially able to reproduce this aspect of natural conditions.

Some time ago we were investigating the cues by means of which mice recognize the position of territory boundaries, and one such experiment involved mounting an enclosure on castors. This enclosure was double-floored as in figure 1.

A perforated floor some four centimetres above the true base of the enclosure helped support a deep layer of sawdust which became extensively tunnelled within a few days. The enclosure at this time contained two territory holders, two subdominants, and four subordinates. A fault in the structure of the enclosure led to a small gap opening in the true floor which was not detected until one morning it was found that some mice had escaped. They were observed before being recaptured and it was found that the two escaped mice were subdominants. One was occupying the floor of the room containing the enclosure and the other the floor of an adjacent room (figure 5). They were each defending their areas against the other and a number of typical territorial encounters were observed on the threshold of the open connecting door. The original territory-holding mice and the subordinates did not leave the enclosure.

It was not practicable to leave the adventitious experiment set up for long but a new enclosure was constructed so that it could be duplicated. This was constructed as in figure 6. The central enclosure was essentially the same as that used in our previous investigations except that it was supported above the floor of the larger pen. It was provided with two small apertures with removable covers which were placed in a position similar to that in which the accidental opening had occurred. Territories were set up in the usual way and when they were established the openings to the outside were uncovered. This experiment was replicated six times and on each occasion where there were recognizable subdominants (4/6ths) they emerged and established their own territories. Three of the primary territory holders extended their defended area to include part of the outside enclosure and one of these gave up his inside space. On

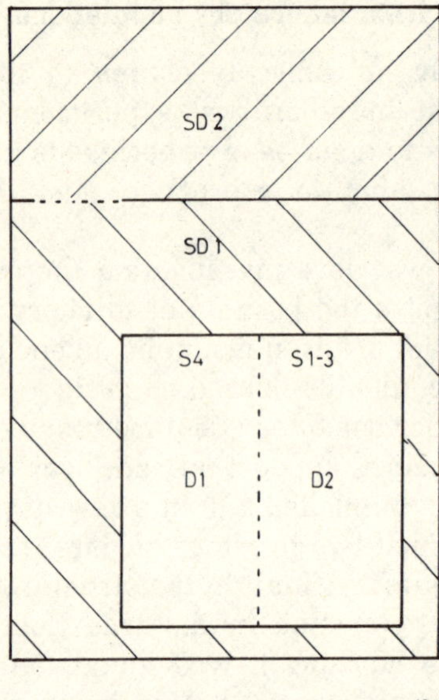

Figure 5 Diagram of territories occupied by escaped mice. D_1 and D_2 are the original territory holders in the enclosure. SD_1 defends the floor of the room containing the enclosure and SD_2 that of an adjoining small room.

those occasions when subdominants were not recognized, one or two subordinates emerged and set up territories. Overall these results were consistent, and show that territory holders are relatively static, whereas subdominants disperse and make territories of their own in unoccupied areas. The role reversal from subdominant to territorial position is in agreement with Crowcroft's finding that even a very badly beaten male can re-establish itself given the opportunity (Crowcroft, 1966). The exact status of the subordinates which dispersed is not known and the possibility remains that they were subdominants with relatively low aggression scores.

It is not unlikely that our subdominants are equivalent to the mobile males described by Berry in wild mouse populations, and dispersal of a similar category individually has been described in voles by Krebs *et al*. (1976). They found that a clear zone was colonized by animals from adjacent populations. The colonists were mostly male, were lighter in weight than those

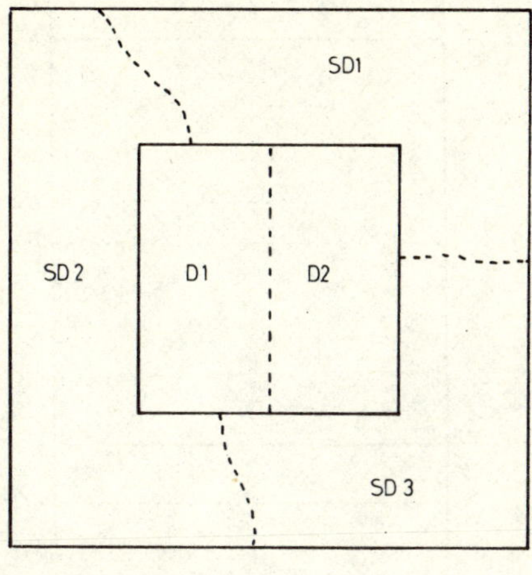

Figure 6 Plan of enclosures used in Dispersal experiments. The dashed line indicates territory boundaries between original territory holders D_1 and D_2 and also between subdominants SD_1, SD_2, and SD_3. (Dispersal experiment 3.)

voles which remained, and also showed genetic differences from the rest of the population.

The frequency with which the four behavioural types occur in our enclosures would suggest that differentiation takes place after a group is established, but may of course be influenced by pre-existing factors.

Effects of available space on aggression

In a previous report it was demonstrated that the position of a boundary between two territories could be manipulated by moving landmarks inside the enclosure (Mackintosh, 1973) and an interaction between space and aggression was found.

If the boundary between two territories was moved in stages so that at each step the territory of one mouse became bigger whereas that of the other was compressed, then the expansionist mouse eventually took over the whole enclosure. This took place before the smaller territory was obliterated by the movement and seemed to be related to a feedback whereby the animal

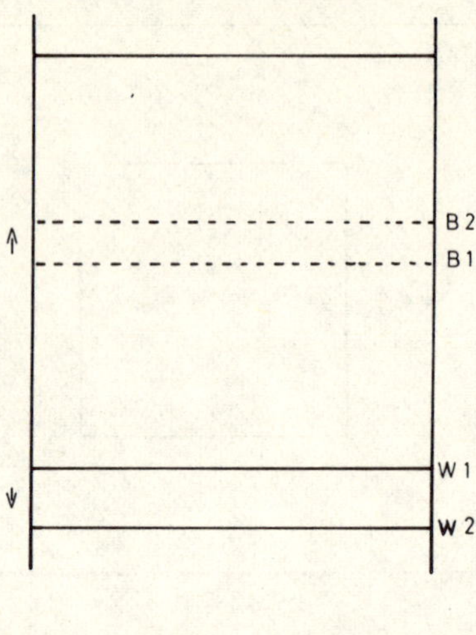

Figure 7 Expandable enclosure. W_1 and B_1 are the initial positions of the wall of the enclosure and the territory boundary respectively. The wall was moved to W_2 and the boundary then changed position to B_2.

given the bigger space became more aggressive and that given the smaller one became less so. This experiment was carried out in a standard enclosure of fixed size and the results have since been confirmed and the inferred relationship between space and aggression supported, using an expandable enclosure.

When the rear wall behind one of the territory holders was moved 30 cm outward as in figure 7, then the territory boundary moved in the opposite direction at the expense of the mouse with the unaltered space. Experiments in which both territories were compressed, on the other hand, had no effect on the position of the territory boundary.

The level of aggression shown by a mouse in these studies therefore varied directly with the amount of space available to it and a similar result was shown in a separate series of experiments concerned with levels of agonistic behaviour.

An enclosure was partitioned as in figure 8 and four mice taken from stock boxes of 20-30 individuals were put into each of the four sections. The frequency of aggressive behaviour was then recorded in hour-long observations made twice a day for a

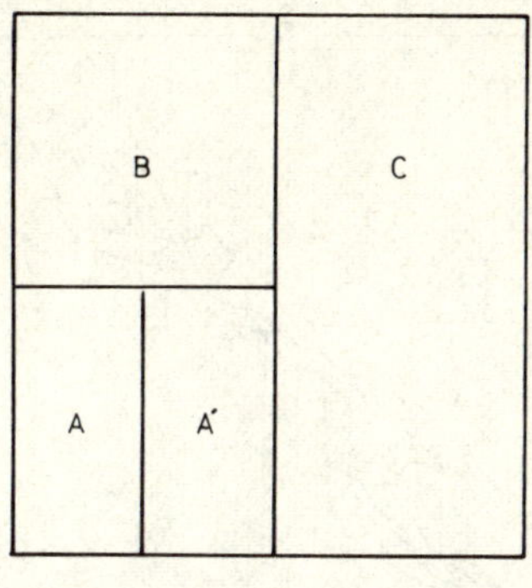

Figure 8 Plan of partitioned enclosure.

week. It was found that the mice in A and A′ showed significantly less aggression than those in C, and that B, with an area of 0.81 m^2, was intermediate.

This area was found to be significant in another and possibly related way. A series of experiments was undertaken to find what was the minimum area of enclosure that could be used to produce a pair of territories. It was found that the limit was 0.81 m^2. The exact dimension at which territory formation becomes impossible may well vary from strain to strain but undoubtedly at about this point the major characteristic of the social behaviour is abolished.

Although we have demonstrated previously that mice use visual information from their surroundings to locate the position of territory boundaries, the gross physical characteristics of the enclosure may be relatively unimportant, as compared with its area. Territories were formed as readily in a circular enclosure as they were in a square one.

The dominance of visual cues is, in any case, not absolute. It was shown previously that in the absence of visual landmarks, olfactory information was used instead (Mackintosh, 1973) and this was followed up by a series of three experiments in which the mice were maintained in total darkness. Precautions were taken so that all light, including transient illumination that might have resulted from opening doors, was excluded from the time

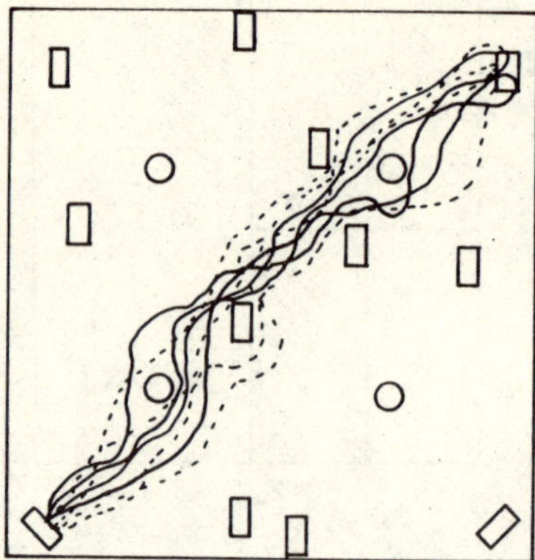

Figure 9 Tracks of two female mice retrieving the same litter in opposite directions.

that the mice were first introduced into the enclosures. Observations were made by means of an infra-red viewing apparatus and in each case territories were established as usual. This result is consistent with those of Stoddard (1970) who compared the breeding performance of two inbred strains in a 12-hour light/dark cycle and in total darkness, and who found that the latter condition had little effect.

The control of populations in laboratory enclosures

When both male and female mice were introduced into our enclosures and the population allowed to increase unchecked the numbers of mice increased until there were approximately 150 per enclosure (Mackintosh, 1970). As was pointed out, this was remarkably similar to the results obtained by Crowcroft (1966) with wild mice, and was equally remarkable for the fact that the method of population limitation appeared to be quite different. Crowcroft reported a diminishing reproductive performance whereas no fall in the production of litters was detected in our colonies.

As far as we could tell the populations were contained at asymptote by the normal retrieving behaviour of the females. Mice tend to retrieve infants when they encounter them (Noirot, 1958) and at high densities litters were encountered very frequently. This meant that the young spent a considerable period being carried. A result of this was that numbers of the young were lost in the sawdust and it is probable that losses also occurred by their being retrieved by non-lactating females. An extreme form of this retrieving behaviour is illustrated in figure 9 which shows the track of two females who carried the same litter backwards and forwards for a considerable period.

Brown (1953) and Southwick (1955) both report infantile mortality as a major factor in the limitation of population growth but both these authors suggest that this results from changes in behaviour with failure of nest building and the appearance of cannibalism.

Territorial behaviour and aggression levels were unaffected by the dense population. Immature mice formed temporary 'bachelor bands' within which there was fighting between the males. Those which emerged as dominants in their peer group formed small territories of their own taking some of the space of the original males. This indicated some compressibility of the territories and confirmed a previous finding that only mice which had established a degree of dominance became territory holders.

Discussion

I have deviated a long way from a discussion of the experimental analysis of overcrowding; however, if the initial proposition that a naturalistic study has more relevance than one in which the artificiality of the situation prevents normal behaviour is accepted, then the experiments with the house mouse provide valuable information.

The evidence shows that the confinement inherent in an experimental laboratory population has many effects. The most significant is perhaps that there is a progressive impoverishment of behaviour with a reduction in available space and that this is operative even in relatively large enclosures. The changes that occur are all likely to alter the response of the animal to high population densities. Even in the largest enclosures, clumping of subordinates occurred and their continued presence in one spot must affect both them and the territorial mice with whom they are in contact. As the space is reduced the next casualty is the site-attachment of females. This is a major loss, as it is unlikely that a population based on permanent pairs or trios will respond in the same way as one in which the females are free to move about. The continual encounters with less familiar males might be expected to result in litter resorption by the Bruce effect (Bruce, 1959; Dominic, 1965), although this was not apparent. The vagrancy of the females, however, certainly contributed to the excessive levels of retrieving behaviour which we found limited our populations. The final reduction in enclosure size obliterated territory formation, which had resisted the high densities of the asymptote populations and the step to the laboratory cage altered the behavioural repertoire.

It must be pointed out that one field study apparently does not agree either with the laboratory results or with the field evidence described above. Newsome (1969) examined a population of house mice in a reed bed in South Australia and he deduced that the ranges of the individual males overlapped extensively and therefore that they were not territorial. He also reported that the population was crowded and that many mice showed bite damage. This, therefore, is a possible case of the breakdown of territorial behaviour under the influence of high densities. At the other extreme Crowcroft (1955b) suggested that territory formation could fail because densities are too low.

Our observations have emphasized the complexity of social structure which mice can demonstrate given the opportunity, so much so that I deliberately adopted the language of primatologists to describe the situation within our dense populations. None of this is visible in laboratory cages where despotic rule by one mouse is the norm.

The independence of available space and density is shown by two points. Firstly, territory formation persisted at high densities but disappeared in small spaces where the density was in fact lower than in the crowded enclosures. Secondly, breeding was controlled in enclosures at a level below that attained by the same number of females in cages.

The elusiveness of the effects of some spatial factors, is illustrated by the work of Swanson (1973) who reported that golden hamsters reproduce very much less in large enclosures than they do in laboratory cages, and indeed showed a zero rate of population increase. This is obviously not a continuing effect as an inverse relationship between breeding rate and available area leads to absurd inferences for the population dynamics of the wild stock.

A large number of workers have reported that levels of aggression rise in dense populations (cf. the review by Archer, 1970) and this behavioural change is often implicated as a controlling factor. Such results have contributed to the popular supposition that density and aggression are positively related and this has been supported by the findings of other investigators who have assembled dense populations directly. The latter case can be misleading as the results are often more the outcome of the animals' pretreatment than of the caging density. Christian (1955) and Bailey (1966) for example, isolated the animals before grouping them and it is well known that isolation produces high levels of aggression in male mice. Experiments of this kind also are usually carried out on unisexual groups and the results of German (1973) suggest that population effects, both on males and females, may be dependent on the density of their own sex rather than that of the opposite one.

The results reported here show that the effect of density on aggression is a complex one and that at some levels these two factors may be inversely related. There was no increase in aggression in our dense populations and two experiments—the expansion of territories and the aggression test in various-sized enclosures—show an increase in aggression with a reduction in density. Both cases are related to territorial behaviour, the former confirming the relationship between territory size and aggressiveness and the latter probably resulting from a triggering of territorial behaviour. The coincidence of size of enclosure between the lower limit for territory formation and the size at which a significant increase in aggression occurred confirms this.

Finally, the initial adrenocortical response of our subordinate mice indicated by organ weight changes is the same as that shown to be produced by high densities (Christian, 1963). No direct measurements were made of hormone levels, but the differences in size of preputial glands would also indicate a suppression of gonadal activity consistent with an increase in adrenocortical activity. Again, however, the situation is complex. Subdominants, which by definition show aggression, are indistinguishable by these means from the most heavily defeated subordinates, as are the category of A subordinates which in fact are not subject to much aggression. Subdominants rapidly assume territorial status when given the opportunity and when strange mice are introduced into a territorial enclosure even B class subordinates have been observed to attack immediately. This suggests a flexibility that is not apparent in cage experiments, just as the rest of the evidence points to the establishment of an increasingly elaborate social structure with increasing approximation to the natural habitat. No overcrowding experiment which is carried out in cages is likely to reflect the way in which mice normally respond to high densities, and care must be used in the interpretation of the results from all confined populations.

References

Anderson, P. K. and Hill, J. L. (1965) *Mus musculus:* experimental induction of territory formation. *Science*, **148**, 1753–1755.

Archer, J. (1970) Effects of population density on behaviour of rodents. In *Social Behaviour in Birds and Mammals*, ed. Crook, J. H. pp. 169–210. New York: Academic Press.

Armitage, B. (1975) Social behaviour and population dynamics of marmots. *Oikos*, **26** (3), 341–354.

Bailey, E. D. (1966) Social interaction as a population regulating mechanism in mice. *Canadian Journal of Zoology*, **44**, 1007–1012.

Barnett, S. A. (1963) *A study in Behaviour*. London: Methuen.

Bergerud, A. T. and Hemus, H. D. (1975) An experimental study of the behaviour of the Blue Grouse (*Dendrogapus obscurus*). I. Differences between the foundings from three populations. *Canadian Journal of Zoology*, **53** (9), 1222–1237.

Berry, R. J. (1968) The ecology of an island population of the house mouse. *Journal of Animal Ecology*, **37**, 445–470.

Berry, R. J. (1970) The natural history of the house mouse. *Field Studies*, **3** (2), 219–262.

Brain, P. F. (1975) Studies on crowding: a critical analysis of the implications of studies on rodents for the human situation. *International Journal of Mental Health*, **4** (3), 15–30.

Brown, J. L. (1969) Territorial behaviour and population regulation in birds. *Wilson Bulletin*, **81**, 293–329.

Brown, R. Z. (1953) Social behaviour, reproduction and population changes in the house mouse. *Ecological Monographs*, **23**, 217–240.

Bruce, H. M. (1959) An exteroceptive block to pregnancy in the mouse. *Nature (London)*, **184**, 105.

Calhoun, J. B. (1961) Determinants of social organisation exemplified in a single population of domestic rats. *Transactions of the New York Academy of Sciences*, Ser. II, **23** (5), 437–442.

Calhoun, J. B. (1962) Population density and social pathology. *Scientific American*, **206** (2), 139–148.

Chance, M. R. A. (1956) Social structure of a colony of *Macaca mulatta*. *British Journal of Animal Behaviour*, **4** (1), 1–13.

Chitty, D. (1952) Mortality among voles (*Microtus agrestis*) at Lake Vyrnwy, Montgomeryshire in 1936–1939. *Philosophical Transactions of the Royal Society, London, B*, **236**, 505–552.

Christian, J. J. (1955) Effects of population size on the adrenal glands and reproductive organs of male mice in populations of fixed size. *American Journal of Physiology*, **182**, 292–300.

Christian, J. J. (1963) Endocrine adaptive mechanisms and the physiological regulation of population growth. In *Physiological Mammalogy, Volume 1*, ed. Meyer, E. and Van Gelder, R. pp. 189–353, London and New York: Academic Press.

Clarke, J. R. (1955) Influence of numbers on reproduction and survival in two experimental vole populations. *Proceedings of the Royal Society, London, B*, **144**, 68–85.

Crook, J. H. (1961) The basis of flock organisation in birds. In *Current Problems in Animal Behaviour*, ed. Thorpe, W. H. and Zangwill, O. L. Cambridge: Cambridge University Press.

Crowcroft, P. (1955a) Territoriality in wild house mice *Mus musculus* L. *Journal of Mammalogy*, **36**, 299–301.

Crowcroft, P. (1955b) Social organisation in wild mouse colonies. *British Journal of Animal Behaviour*, **3**, 1–36.

Crowcroft, P. (1966) *Mice All Over*. London: Foulis.

Crowcroft, P. and Rowe, F. P. (1963) Social organisation and territorial behaviour in the wild house mouse (*Mus musculus* L.) *Proceedings of the Zoological Society of London*, **140** (3), 517–531.

Dominic, C. J. (1965) The origins of the pheromones causing pregnancy block in mice. *Journal of Reproduction and Fertility*, **10**, 469–472.

Eibl-Eibesfeldt, I. (1950) Beiträge zur Biologie der Haus- und der Ahrenmaus nebst einigen Beobachten an anderen Nagern. *Zeitschrift für Tierpsychologie*, **7**, 558–587.

Elton, C. (1942) *Voles, Mice and Lemmings*. Oxford: Clarendon Press.

Evans, C. M. and Mackintosh, J. H. (1976) Endocrine correlates of territorial and subordinate behaviour in groups of male CFW mice under semi-natural conditions. *Journal of Endocrinology*, **71**, 91.

German, A. L. (1973) Effects of sex ratio in an experimental population of laboratory mice on the weight of individuals. *Ekologika*, **4** (5), 96–98.

Grant, E. C. and Mackintosh, J. H. (1963) A comparison of the social postures of some common laboratory rodents. *Behaviour*, **XXI**, 3–4.

Hediger, H. (1950) *Wild Animals in Captivity: An Outline of the Biology of Zoological Gardens*. London: Butterworths.

Henry, J. P., Ely, D. L., Watson, F. M. C., and Stephens, P. M. (1975) Ethological methods as applied to the measurement of emotion. In *Emotions—Their Parameters and Measurement*, ed. Levi, L. pp. 469–487. New York: Raven Press.

Hinde, R. A. (1956) The biological significance of territories in birds. *Ibis*, **98**, 340–369.

Krebs, C. J., Wingate, I., Leduc, J., Redfield, J. A., Taitt, M., and Hilborn, R. (1976) Microtus population biology. Dispersal in fluctuating populations of *M. Townsendii*. *Canadian Journal of Zoology*, **54** (1), 79–95.

Mackintosh, J. H. (1970) Territory formation by laboratory mice. *Animal Behaviour*, **18**, 177–183.

Mackintosh, J. H. (1973) Factors affecting the recognition of territory boundaries by mice (*Mus musculus*). *Animal Behaviour*, **21**, 464–470.

McBride, G. (1971) Theories of animal spacing: the role of flight, fight and social distance. In *Behaviour and Environment. The use of Space by Animals and Men*, ed. Esser, A. H. New York: Plenum Press.

Newsome, A. E. (1969) A population study of house-mice permanently inhabiting a reed-bed in South Australia. *Journal of Animal Ecology*, **38**, 361–377.

Noirot, E. (1958) Analyse du comportement dite maternelle chez la sourie. *Monographie Francaise de Psychologie*, No. 1. CNRS, Paris.

Oortmerssen G. A. van (1970) Biological significance, genetics and evolutionary origin of variability in behaviour within inbred strains of mice (*Mus musculus*). *Behaviour*, **38**, 1–92.

Poole, T. B. and Morgan, H. D. R. (1976) Social and territorial behaviour of laboratory mice (*Mus musculus* L.) in small complex areas. *Animal Behaviour*, **24** (2), 476–480.

Reimer, J. D. and Petras, M. L. (1967) Breeding structure of the house mouse (*Mus musculus*) in a population cage. *Journal of Mammalogy*, **48**, 88–99.

Scott, J. P. (1944) Social behaviour, range and territoriality in domestic mice. *Proceedings of the Indiana Academy of Sciences*, **53**, 188–195.

Southwick, C. H. (1955) Regulatory mechanisms of house mouse populations: social behaviour affecting litter survival. *Ecology*, **36**, 627–634.

Stoddard, R. C. (1970) Breeding and growth of laboratory mice in darkness. *Laboratory Animals*, **4**, 13–16.

Swanson, H. (1973) The consequences of overpopulation. *New Scientist*, 26 July 190–192.

The dynamics of spatial behaviour

L. R. TAYLOR
Rothamsted Experimental Station, Harpenden, England
and
R. A. J. TAYLOR
Imperial College Field Station, Silwood Park, Ascot, England

This paper considers the proposition that mobility provides an alternative to competition in the control of populations. It deals with concepts at population level and because it is difficult to identify the effects, in a population, of the behaviour of an individual, it argues from what is known of populations to what we suppose is the cause.

We equate behaviour with motion and classify all motion into three sets, feeding, reproductive, and distributive. The third component, which we call spatial, is made responsible for changes of abode and, by definition, the first two kinds of behaviour then constitute motion that is random with respect to the third.

In nature all organisms, from protozoons to man, are shown to have a common element of spatial distribution that is strongly dependent on density and we deduce that the spatial behaviour which leads to these distributions must itself be density dependent. This density-dependent spatial behaviour is further classified as that which increases the distance between individuals and that which decreases it; studies on aphids illustrate some of this behaviour.

Power functions are then used to model such a process and the resulting population patterns match those found in nature; they also appear to produce intrinsic population control.

We conclude that the mechanism proposed is viable, in principle if not in detail, but emphasize that the model is of negative competition, not altruism nor cooperation, although only the spatial outcome, not the behavioural motivation, is considered.

The problem

The central theme both of population dynamics and evolution is that populations build up, the environment changes, and populations disappear. If a population builds up solely by births and declines only by deaths, the process conforms to the classical population dynamics of Malthus, Verhulst, and Darwin, in which maximized reproduction is limited by extrinsic mortality, that is, mortality imposed by external factors. But if the population assembles by movement and then disperses, explanations are sought in intrinsic animal behaviour. Nevertheless, both scenarios depict the rise and fall of a population and therefore have much in common.

As it stands, the first scenario would seem to be unrealistic because populations cannot survive through many generations without some movement. At least the founders of a population must be immigrants and other populations must produce emigrants to provide these. This means that classical population dynamics do not deal with living, moving, individuals but with objects that appear and disappear without volition; they have no intrinsic motion. In other words, Malthusian dynamics are based on a physical, not a biological, model. To make the model biological, individuals should be given the ability to choose from alternative courses of action; they should behave.

For some time it has been recognized that, if animal behaviour could be integrated with population dynamics, the density-dependent negative feedback that is essential to maintain population control could come from intrinsic behaviour instead of extrinsic mortality (e.g. Grinnell, 1904; Skellam, 1951; Lidicker, 1962), although it has perhaps not been made quite so explicit as this. However, the fundamental issue then at stake is whether or not fitness would be impaired. Does the emigrant that makes possible the founding of a new population lose fitness in so doing? In reference to Grinnell's (1904) belief that there is a centrifugal flow of individuals from population centres, MacArthur (1972) perceived such a loss of fitness unless the individuals return at another season.

There is now sufficient evidence to present an alternative, giving individuals the choice to avoid competition without loss of fitness, by moving. We suggest that the confidence placed in positive competition as the sole directive force in evolution is too great and that the avoidance of conflict by movement—negative competition (Hutchinson 1957)—is equally important. Certainly, if movement is an acknowledged component of population dynamics, fitness can no longer be defined simply as 'the number of adult offspring left in the next generation' (e.g. West Eberhard, 1975) because, lacking volition, these offspring would be dispersed at random by the physical environment and we know that spatial randomness is biologically rare (Taylor and Woiwod, 1978). We suggest that the clues about population density received by the individual through social behaviour *per se* (Wynne-Edwards, 1962; Wilson, 1975) activate negative feedback by spatial behaviour represented in population dynamics as movement, rather than reproductive restraint.

Kennedy (1969) defines behaviour as 'the integrated functioning of the whole animal in its environment with special reference to its movements'. For present purposes we would simply say that behaviour is self-induced motion. When a whole organism turns sensory information into action it behaves, and any resulting motion produces patterns of individuals in space. These patterns are specific and well known for many organisms. Whether an individual moves from or stays in, a given place depends on the internal and external environmental pressures exerted on it. We suppose that for each individual, successful survival and reproduction consists in 'learning' when to move and when not to move so as always to have the best prospect of occupying the optimum environment and in transmitting this accumulated learning down the generations. The ecological motivation was crystallized, although perhaps overstated, by Preston (1969) when he wrote 'Most animals are mobile for the express purpose of changing their local density. The grass is generally greener on the other side of the fence'.

The population constellation

When an individual's life history is represented in a spatial plane, latitude and longtitude without time, a modified Cavalli-Sforza (1962) loop diagram (figure 1) shows the three classes of behaviour we mentioned earlier. Feeding behaviour and mating behaviour appear as loops beginning and ending at the same coordinates, F, M_1, M_2, O_1, O_2, . . . In contrast, the intervening vector quantities, BO_1, O_1O_2, . . ., involve a lasting change in spatial coordinates. The three fundamental sets of behaviour

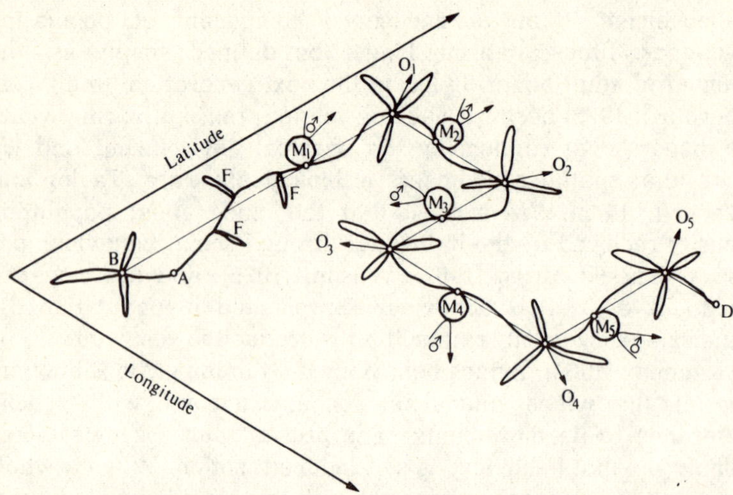

Figure 1 Diagrammatic representation of a generalized life-history without time. B = birth; F = feeding loop; A = adulthood; M = mating loop, O = offspring; D = death. By definition, feeding and mating loops are 'trivial' behaviour and do not contribute to change of abode represented by the thin vector line. Modified after Cavalli-Sforza (1962).

can now be reduced to two which are recognizable as, but not identical with, what entomologists call 'trivial' and 'migratory' behaviour (Heape, 1931; Southwood, 1962), where to migrate is defined as 'to change one's abode' (Oxford English Dictionary). Unfortunately, segregation is not always so simple in practice as it appears in this diagram although it seems that all the essential elements are represented there; we should perhaps emphasize that we cannot identify, quantify, and give reference to all the components of spatial behaviour for all species, any more than one could specify all the different kinds of feeding behaviour in a general model for ingestion. Each species has highly specific methods of spacing and their only common component is in the resulting spatial pattern of the population.

By eliminating the 'trivial' components of behaviour and restoring time, the model is greatly simplified to yield Hägerstrand's (1962) individual life-lines (figure 2), the vectors BO, O_1O_2, O_2O_3, . . . , which touch at mating, divide at reproduction, and end at death. Integrated over all the adults in

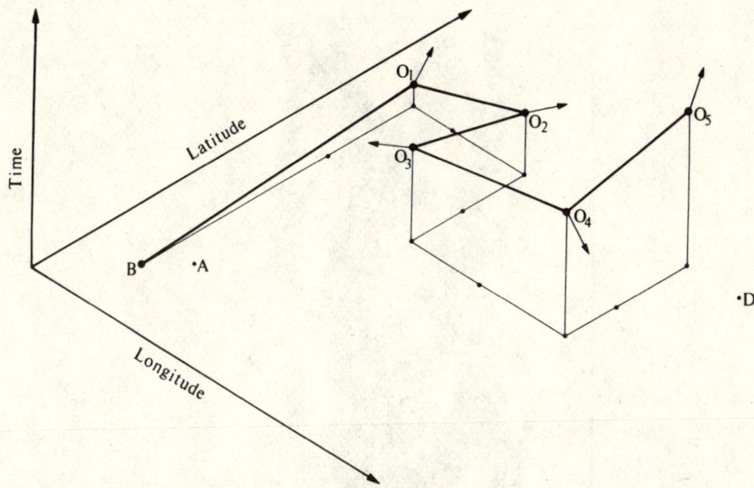

Figure 2 Generalized life-line from figure 1 with time introduced. Symbols as in figure 1. Trivial movements are eliminated to leave only the change-of-abode vectors in three dimensions connecting the birth places of the generations (O becomes B in the next generation). Modified after Hägerstrand (1962).

a population the resulting skein of vectors expresses in three dimensions all the behaviour relevant to population dynamics and it may be presented by the coordinates at B, O_1, O_2, . . . , as a constellation in space–time (figure 3). Horizontal sections through this constellation of individuals are the density–distribution maps of real populations (figure 4), and vertical sections are the density changes through time along a transect (figure 5). The separate strands in figure 3 may be regarded as population elements, colonies, demes, groups, or just as local densities, such as are sampled to make a map like figure 4, but there are no geographically fixed limits to them. Individuals can cross the gaps between.

A frequency distribution of numbers of individuals in these population elements forms a regular, often hollow, curve (figure 6). There is a vast and inconclusive literature about these frequency distributions, but they all have one thing in common: no matter what the organism, protozoon, worm, insect, mollusc, echinoderm, fish, bird, or mammal—not excepting man—they have a characteristic spatial variance proportional to a frac-

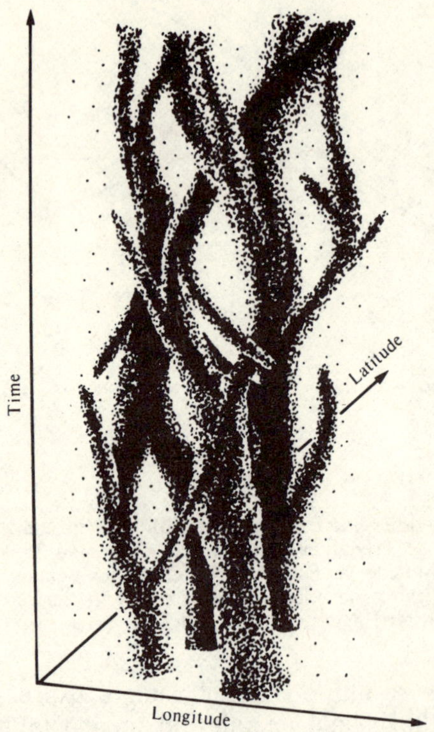

Figure 3 When all the individuals of a species are represented in space-time by the points B (O) in figure 2, the resulting three-dimensional scatter is seen as an anastomozing constellation with strands representing populations, demes, or local densities. Modified after Taylor and Taylor (1977).

tional power of mean population density (Taylor, 1961, 1971). If a good field naturalist should happen to find some of the members of a population, he will probably be able to find more of them because he intuitively recognizes this spatial characteristic, some species being highly aggregated and others more evenly spaced. Because the relation between spatial variance and population density is a power function with an index averaging more than one (1.48 ± 0.39) (see figure 7), spatial behaviour must change as density changes, to produce more extremes of concentration and isolation of individuals at high densities than could be forecast from low densities by simple

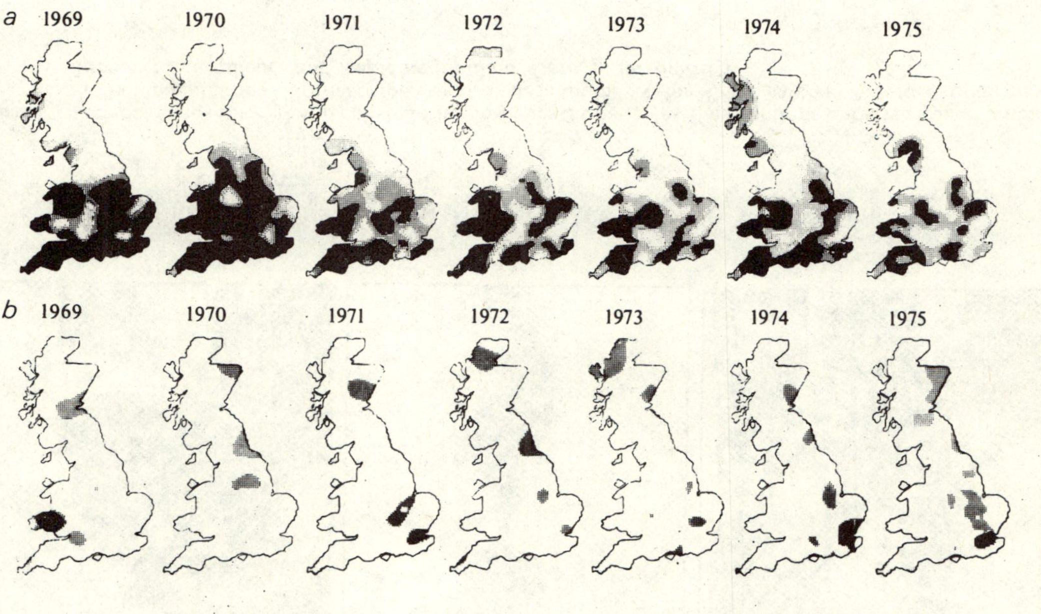

Figure 4 Horizontal sections through the constellation in figure 3 are seen in real populations as changing maps of the distribution of density: (*a*) the buff ermine moth (*Spilosoma lutea*) has a fairly solid population structure with moving 'holes'; (*b*) the garden dart moth (*Euxoa nigricans*) has a population of isolated strands, only occasionally anastomozing. Data from 60–109 sampling stations of the Rothamsted Insect Survey; layering intervals are logarithmic (Taylor, 1974).

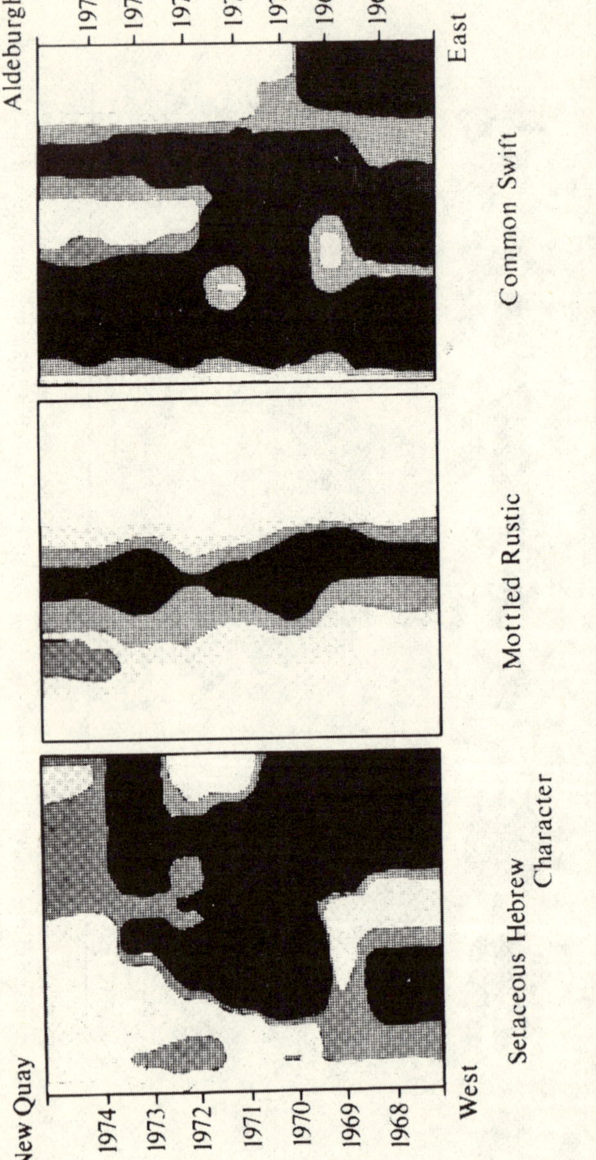

Figure 5 Vertical slices through the constellation in figure 3 appear as time-sections of density distribution in real populations. Transect of Great Britain at 52°N for three moths, *Amathes c-nigrum*, *Caradrina morpheus*, and *Hepialus lupulina*. Data as for figure 4.

proportion (Taylor and Woiwod, 1978). We regard this as convincing evidence that spatial behaviour is density dependent.

The spacing also reflects the heterogeneity of the environment which is unique for each individual and different from that experienced by others in the same terrain because each one responds to only a part of the total environment. (For example, a cat sees none of the colour pattern so important to a bird which, in turn, knows nothing of the world of scents inhabited by a dog.) Spatial heterogeneity thus consists of the individual's own sensory impressions of its food, enemies, diseases, and other members of its own species, as well as of the topography and climate, and is therefore constantly changing, especially

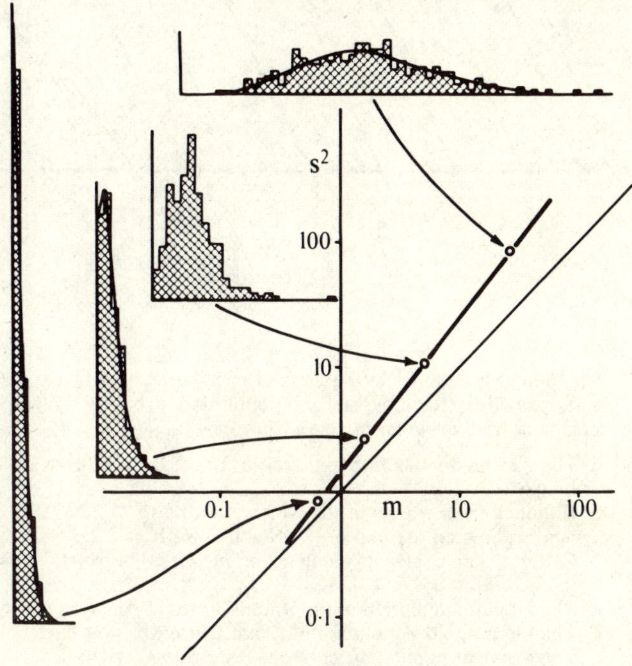

Figure 6 Frequency distributions of the population densities in demes, or the strands of figure 3, or per unit area in figure 5, change disproportionately to give more extremes of isolation and concentration at higher population levels than expected by simple proportion. Spatial variance is a power function of mean density; $s^2 = 1.5m^{1.25}$ for the corn earworm i.e. spatial variance is density dependent. Data from McGuire, Brindley and Bancroft (1957), after Taylor (1971).

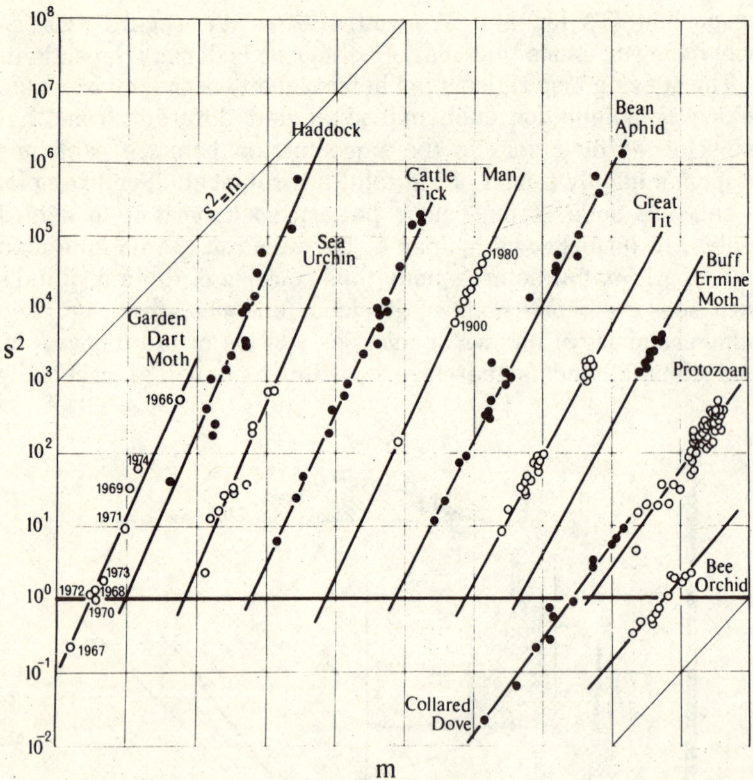

Figure 7 The power function relating population variance (s^2) to mean density (m), provides the only spatial population property known to be common to all organisms from protozoons to man.

A The garden dart moth population of Great Britain (b = 2.36) (see figure 3).
B Haddock from trawls in the North Atlantic (b = 2.33).
C Sea urchins on the bed of the North Sea (b = 2.27).
D Cattle ticks in sweep samples of pastures in North Australia (b = 2.16).
E The human population of the United States of America (b = 2.04).
F The bean aphid population of Great Britain (b = 1.88).
G The great tit population of Great Britain (b = 1.79).
H The buff ermine moth population of Great Britain (b = 1.71) (see figure 3).
I The collared dove population of Great Britain (b = 1.32).
J A ciliate protozoon population on flatworms in North Wales (b = 1.31).
K A bee orchid population in South England (b = 1.15).

Examples from 131 sets of data in which the mean power (b) is about 1.5, showing strong density dependence of spatial behaviour. Modified after Taylor and Woiwod (1978).

with population density, its own age, and its physiological condition.

Because individuals of the same species respond mainly to the same components of the environment, we suppose that the behaviour of others may be a valuable clue to the distribution of resources and enemies, and of other advantages and disadvantages. Where others are surviving, survival must be possible and this suggests an advantage in joining them; but also, where there are others, resources have been consumed and enemies may have congregated, reducing the initial advantage and suggesting a move away. We suppose that throughout evolution this experience of other individuals as environmental clues has been reflected in the individual's responses to each other and so has built up the density dependence we find in spatial behaviour. This spatial behaviour can all be classified into two mutually exclusive categories. Attraction often, but not always, appears to be sexual and we suppose predominates at low density and tends to increase it; repulsion is perhaps seen mainly as overt competition, which prevails when density is high and tends to decrease it. Consequently there is a density at which these pressures are in balance but, in any locality, the balance will be upset repeatedly by reproduction and may then be restored by movement activated by the individual's own, or its offspring's, interests.

The spatial cohesion of a population, then, is seen as being maintained by the responses of one individual to another either directly or through some common interest in the environment, but the response may not be immediate; it may be delayed even until a later generation. It is necessary therefore to show an example of how subtle and flexible these responses can be. To do this we need to select a highly mobile organism in which the temporal and spatial scales show change that is rapid compared with our own. Also, because we are concerned with migration, migrants should be readily recognizable.

This symposium is mainly concerned with homoiothermic vertebrates and so lays emphasis on sexual reproduction and behaviour. Feeding behaviour is often complex and confounded with reproductive behaviour, in territorial possession for example, and long, complex, life-histories confuse the life-lines. We therefore look for guidance to highly migrant organisms in which reproductive behaviour concerned with sex and all its elaborations is limited and in which feeding behaviour is simplified so that Cavalli–Sforza loops are minimized and Hägerstrand life-lines are clearly exposed.

Spatial organization in the Insecta

Most insects are small and winged and use their wings mainly for population redistribution (Johnson, 1969). Many use their wings only once, to migrate. Migrants are often easily recognizable. Many highly successful insects, if success is judged by numbers of individuals or species, have evolved asexual life-styles secondarily and, in their clonal populations, reproductive behaviour is minimal and genotypic and phenotypic variability are clearly segregated. In his original specification for the role of competitive exclusion in the abstract niche, Hutchinson (1965) singled out the Homoptera as particularly intractable because of their spatial versatility and we now turn to population structure in this group to examine spatial behaviour, in particular to the Aphididae, a highly successful taxon of about 4000 species, all of which have an asexual stage in their population cycle. We shall see how much behavioural control a clonal population can exert over its own spatial dynamics through mechanisms that respond to environmental cues, many of which reflect the pressure of other individuals. The evidence for the spatial behaviour of aphids comes from a range of species so that the picture presented is a mosaic.

Aphids go through alternating cycles of population growth and migration and usually also of sexual and asexual forms (figure 8). The wingless multiplicative morphs are neotenous and depend on the environment for the suppression of wing production, as also do the asexual forms for the suppression of the sexual condition. A typical example, the peach-potato aphid *Myzus persicae,* goes through the sexual cycle in winter in Central Europe but in Great Britain this is uncommon, the species being largely androcyclic, producing a few males but only virginoparous females in three seasonal cycles of population growth and dispersal, each of which faces different environmental problems (Blackman, 1974; Taylor, 1977). In an adaptation to the tropics the sexual cycle is usually absent although

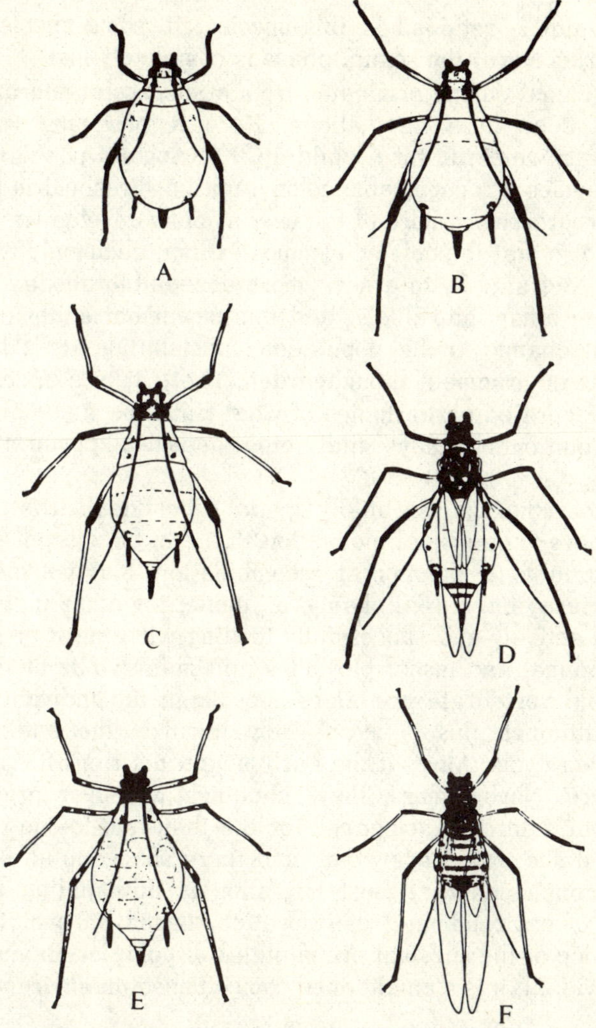

Figure 8 The polymorphs of an aphid (*Megoura viciae*) identify migratory status as well as sex and reproductive capacity.

A First generation virginoparous viviparous apterous female; the fundatrix.

B Second generation virginoparous viviparous apterous female; the fundatrigenia.

C Apterous virginoparous vivaparous female from long-established asexual line.

D Alate virginoparous viviparous female.

E Apterous oviparous, sexual female.

F Alate male.

After Lees (1961).

its potential is retained in this species. In some species, like *Aphis craccivora*, the sexual phase is completely lost.

Each clonal colony originates from a successful migrant that arrives alone in a new abode. Here it produces wingless individuals which, in turn, build up the wingless population, in many species a recognizable colony, until a threshold is passed when progressively more of the new nymphs develop wings and emigrate so that the colony becomes extinct, commonly within a season. Migrants fly to a new, more acceptable abode, usually free from other individuals, and this movement epitomizes the constant change in the population constellation, tracking the changing environment (Roughgarden, 1974). It is a speeded-up and exaggerated performance of what Hutchinson (1965) called 'the evolutionary play' that other species perform more sedately.

The movement is essentially spatial adjustment. It is present when there is no sexual motivation that can be recognized and involves no gene selection or recombination. It is not merely a feeding foray but a real change of abode, for other individuals are still actively and successfully feeding when most emigrants leave. Fungal and insect parasites and many kinds of invertebrate and vertebrate predators may claim the individual if it stays, although this is more likely towards the end of the population cycle. Most of the early emigrants run little risk on this score. Nevertheless, they relinquish whatever protection the colony affords in exchange for the thousand-to-one chance against a successful migration, and there seems no alternative to the conclusion that most migration is anticipation by the species of environmental deterioration, based on evolutionary experience of the survival probabilities of going or staying. For the individual, it is a conditioned choice based on environmental stimuli.

Feeding behaviour in aphids is minimal because they are essentially static once their stylets have found the right food supply. The host has been selected by the immigrant progenitor and the progeny only require to locate the appropriate tissue, usually the phloem, to tap the nutrient reservoir on which they stand; that is, unless the immigrant has made a mistake which, as we shall see later, does sometimes happen. In species such as the bean aphid, *Aphis fabae*, nymphs commonly settle close to their place of birth. This is not because the parent supports the young, which are quite self-sufficient, but there is an element of active congregation between individuals (Ibbotson and Kennedy, 1951), which is evident in the statistical distribution of individuals in space (Taylor, 1970). In several species, immigrant, parthenogenetic alatae that arrive separately on the same plant, presumably from different clones, congregate (Kennedy and Crawley, 1967) so that some colonies are mixed clones. This active congregation may have protective value against enemies but it also seems possible that the earlier individuals provide a visual cue for later arrivals that identifies good feeding sites. Whatever are the origins of the several elements of behaviour that cause aggregation, they are not merely sexual and they fulfil a function of increasing local density.

Control of migration in aphids

The development of migrant polymorphism and migratoriness leads to a sophisticated and sensitive spatial control system. Four major environmental factors, food, crowding, photoperiod, and temperature, control the formation of wings. In addition to the production of these winged migrants, their willingness to migrate, given the ability, is activated or inhibited by other factors such as the physical and chemical features of the take-off or alighting surface, light intensity, and wind speed.

Food responds actively to the density of the population feeding. In some instances feeding improves the palatability of the host plant by generating galls (Kennedy, 1958). In others, the plant loses turgor pressure and in a few species, such as the greenbug (*Schizaphis graminum*), toxins are injected during feeding and may kill the plant (Wadley, 1931). These reactions in turn affect subsequent morph determination (Forrest, 1970), so feeding may produce positive or negative feedback in morph control that is an indirect response to the other individuals in the constellation.

The crowding effect, *'l'effet de groupe'* of Bonnemaison (1951), controls morph production directly in response to other individuals. The primary sensory mechanism is tactile (Johnson,

1965) with thresholds related to the level of activity of the individual in *Acyrthosiphum pisum* (Lowe and Taylor, 1964). It can be induced by contact with other small insects but it is more usually induced by other members of the colony (Lees, 1967). The sensitive period may be during the parent's early life, its adult life, or during the early life of the migrant itself, or a combination of these depending on the species (Shaw, 1970a). The tactile receptors are on the antennae and elsewhere, and removal of the higher nervous centres ensures uninhibited migrant production (Lees, 1975). Food quality can over-rule previous tactile experience, the production of presumptive migrants sometimes being reversed when artificial food is replaced by fresh young leaves (Johnson and Birks, 1960).

Sex determination is interlinked with migrant control because suppression of males and oviparous females leads to the virginoparous condition, and hence to further potential restriction of wing production, so that photoperiodism is indirectly involved in migration. The *pars intercerebralis* inhibits sex-production by responding to the length of the nocturnal dark period interrupting daylight (Lees, 1964). In the absence of the inhibitors, normal bisexuality is restored.

Temperature affects wing production directly, but its side effects are so complex in poikilotherms that it is almost impossible to disentangle them completely.

Of more concern here is that release from the sexual inhibition can be delayed by an intrinsic timing mechanism (Lees, 1960), the *facteur fondatrice*, which may prevent the generation of sexual forms for a time period, of up to several months in one species (*Drepanosiphum platanoidis*) (Dixon, 1971), irrespective of the number of generations passed through during the latent period, at least in *Megoura viciae* (Lees, 1961). A similar inhibitor–retention mechanism prohibits the production of emigrants directly from immigrants, delaying the onset of the migrant phase.

Beyond this fairly superficial level of migrant control by inhibition of wing dimorphism lies what Lees (1975) has called a 'bewildering array' of less obvious phenotypes which extends the behavioural control of migration and reproduction into a more subtle spatial control system.

Depending on internal and external conditions, winged individuals of *Aphis fabae* and other species may be compulsive migrants, incapable of feeding and reproducing without first taking flight, or being subjected to some artificial substitute such as enforced walking (Johnson, 1958). Others are capable of flight but can be induced to indulge in it only briefly and by adverse conditions (Dry and Taylor, 1970; Shaw, 1970b). A third condition is winged but lacks wing muscles and a fourth has all the various gradations of wing abortion. Wingless morphs again may be more or less migratory, although migration must then be accomplished on foot, and they bear morphological markings that indicate their migratory status (Shaw, 1970c). The most important activating factor here is population density, acting through direct contact between individuals. The series forms a complete spectrum of migratory response to the whole environment, especially the social and internal environments, acting over a period of many generations, so that the density dependence of spatial behaviour may be delayed, the offspring migrating in response to the density of previous generations.

In addition to the migratory condition of the migrants, initiation of take-off (Dry and Taylor, 1970), which is an immediate response to the colour and intensity of light (Moericke, 1950), and of flight itself (Halgren and Taylor, 1968), are both controlled by temperature and light thresholds (Taylor, 1957) so that the onset of adverse conditions can inhibit migration, predetermined from generations back, until the last moment. Or once started, migration can be stopped and there is a progressive change in the balance between induction and depression of flight activity (Kennedy, 1966). The final limit to the migratory ambit is set by the reversal of the light response (Kennedy, 1966) and the flight thresholds of light and temperature, combined with the onset of wing muscle autolysis (Johnson, 1957, 1959), which restrict migration to one daylight period or permit flight through the night, depending on the climatic region (Berry and Taylor, 1968). Sufficiently long inhibition of flight at this stage, by darkness or even by mutilation of the wings (Chiang, 1960), may reverse the cycle so that the migratory compulsion is lost and the aphid returns to feeding and reproducing.

The resulting wide range of behaviour creates a mechanism for scanning the environment for suitable sites at every scale, from a leaf to a continent, by different kinds of migration on different occasions (Taylor, 1965). Having found a suitable site to establish a colony, the reproductive processes cover an almost equally bewildering array of possibilities in which, although the actual mechanisms are controversial and depend upon the species (Dixon, 1976), individual size, which is a response to temperature, nutrition, and other individuals, can operate as a polymorphic mechanism to control population growth rates, and hence the next cycle of migration (Taylor, 1975).

All this intricate mechanism, and the control it exerts over behaviour and hence the phenology and spatial distribution of populations, is phenotypic, retained over thousands of generations in clonal cultures and correspondingly far removed from direct gene control.

The significance of aphid migration

Why should an aphid risk migration when it does? Is it all selfless destruction for the good of the clone or kin (Hamilton, 1972) and hence peculiar to the asexual condition? This seems improbable because many of these migratory behaviour patterns are found in the sexual stages as well as in the virginoparae. Similar compulsive migratory behaviour also occurs in Diptera, Lepidoptera, and Coleoptera, and many other insects that have a normal bisexual, egg-laying population. Most insect migratory behaviour has much in common with it (Johnson, 1969). In some species of Homoptera, such as *Cicadulina*, a genetic connection has been demonstrated (Rose, 1972). Aphids are merely extreme exemplars of this way of life in which the asexual component is not an essential, but an additional, specialization that we have used to illustrate the point that movement in relation to other individuals may be quite independent of sexual motivation. Aphids do not always stay to compete with other aphids, predators, and parasites, nor to eat themselves out of resources; they move on before these things happen. Movement is thus a behavioural strategic alternative to competition, using both words in a general sense.

We see spatial behaviour as a balance between staying to compete and leaving to explore; it constitutes a choice exercised by the individual, given the limitations of its history and environment, that offers alternative prospects for survival. Aphids cover many spatial options with different individuals each making a choice that may become effective only in future generations. The behavioural polymorphism so readily visible in their migratory morphology points up the deficiency in a numerical concept of fitness based only on the number of adult offspring in the next generation; their geographical behaviour is at least as important as their numbers.

The necessity to be in the right place at the right time is sometimes very clearly illustrated by aphids because any folly of inappropriate spatial behaviour becomes magnified during the population growth of the static phase only to flare up again at the next migration. In a near relative of *Aphis fabae*, the Tropaeolum aphid, migratory behaviour is not matched to feeding and reproductive behaviour. When wingless generations have built up a large, apparently successful population on one host, new adults emigrate and persist in becoming resident on hosts where they can survive to reproduce—but their young invariably die (Taylor, 1959). The lack of fitness of these migrants consists almost entirely in migrating to the wrong place. High fertility is then of secondary importance.

Aphid migration is usually so obviously successful that it cannot seriously be regarded as involving a loss of fitness. Although the behaviour may be less obvious in individuals of other species, all individuals who do not stay to compete with their own parents for the abode they were born in, are following the same dynamic pattern as the aphid, even when they move only a short distance inside the area known to be occupied by the population. If a bird lays its eggs elsewhere than in the nest it was born in, it too has migrated in the sense used here. This usage of the word migration—to change one's abode—for the most potent element of spatial behaviour, has been chosen deliberately to emphasize the positive aspect of spatial behaviour.

The out-and-home character of migration, as used in ornithology for example, classifies some species as non-migrants. Their spatial behaviour is then called 'dispersive' with its unfortunate implications of fixed population centres and random losses, as in the sedentary European blackbird (figure 9), for

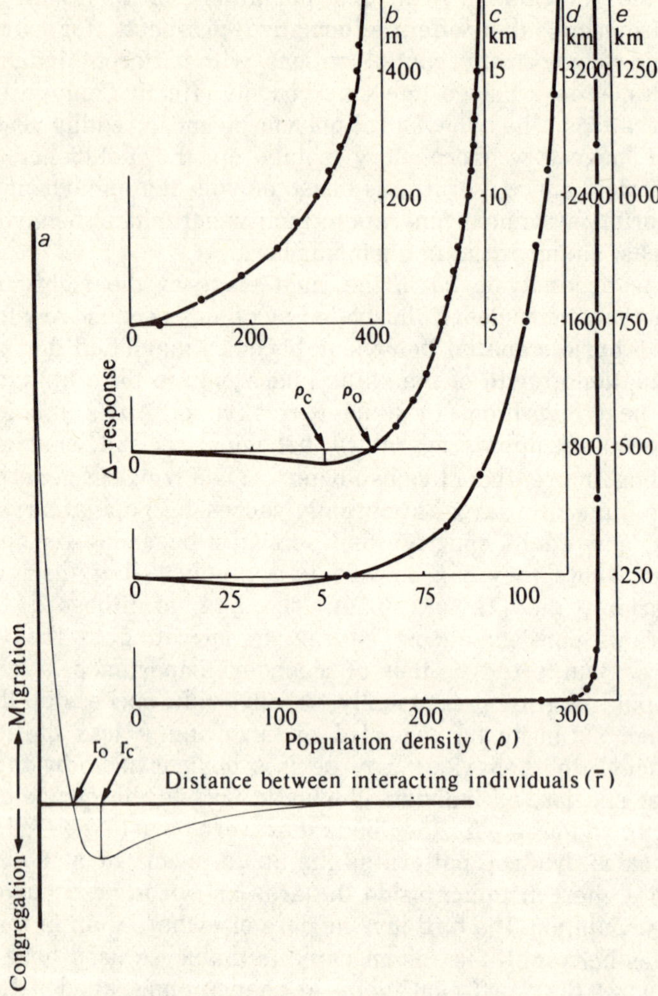

Figure 9 The Δ-function (*a*) gives the relation between displacement and the mean separation between interacting individuals. Converted to density × distance, the function describes data from: (*b*) a sedentary insect (*Drosophila pseudoobscura*), (*c*) a social primate (man), (*d*) a migrant insect (monarch butterfly), (*e*) a territorial bird (European blackbird). Mean separation ($\bar{r}$) is proportional to $\rho^{-\frac{1}{2}}$. At ρ_0 migration and congregation are in balance and the population is spatially stable. At $\rho > \rho_0$ population is over-dense and migration ensues; at densities between ρ_c and ρ_0 congregation rapidly restores balance; below ρ_c spatial recovery is greatly impeded. After Flowers and Mendoza (1970) and Taylor and Taylor (1977).

example. We wish to draw attention to the high fitness of these movements in a continuum extending from great distances down to local movements without recognizing boundaries of demes or populations.

Spatial behaviour modelled

Earlier we argued that all spatial dispositions can be regarded as resulting from the balance between two antithetical sets of spatial behaviour between individuals; repulsion behaviour which diminishes local density by centrifugal movement or migration, and attraction behaviour which increases local density by centripetal movement or congregation.

Elsewhere (Taylor and Taylor, 1977) we have postulated that the net displacement of an individual (Δ) is proportional to the difference between two fractional powers of the population density (ρ):

$$\Delta = \epsilon\left\{\left(\frac{\rho}{\rho_0}\right)^p - \left(\frac{\rho}{\rho_0}\right)^q\right\}, \tag{1}$$

where ρ_0 is the biologically identifiable parameter giving the density at which migration exactly balances congregation, and ϵ is a scale factor containing the dimensional units. Congregation predominates when $\rho<\rho_0$ and migration when $\rho > \rho_0$ (figure 9). The exponents p and q are rate constants for density-dependent migration and congregation respectively. In figure 9*a* the Δ-function is drawn against, $\bar{r}$, the mean separation between individuals ($\bar{r} \propto \rho^{-\frac{1}{2}}$). It shows the theoretical migratory response predominating when individuals are close together, giving way to the congregatory response as the individuals move further apart. Figures 9*b*, *c*, *d*, and *e* show the more familiar hollow curves of distribution of density in relation to distance travelled (Taylor, 1978) that results from the Δ-response in several widely different species. There are no data for the bottom end of these curves, perhaps because it has not seemed of interest, but also because it is difficult to obtain measurements there.

To investigate these effects of population redistribution we have simulated populations in which each individual moves according to the Δ-function (equation 1) on an environmental grid where each space is allocated a reproductive category: Benign, which permits reproduction; Tolerant, permitting survival but not reproduction; and Hostile which kills the immigrant, by whatever means, before it can reproduce. The categories are allocated basically at random and can be re-allocated between generations. If redistribution of the population according to the Δ-function follows after each reproduction by binary fission, to simulate a primitive condition early in evolution say, the population reaches a quasi-stable spatial pattern, in part determined by the parameters of the $\bar{\Delta}$-function and partly by the pattern of heterogeneity of the environment matrix of 130 × 120 elements. In most trials, stability was reached at $\rho \sim \rho_0$ after a very few generations. The total number in the population (N_t) varied very little once numerical stability had been reached, no matter how many generations were followed, although redistribution continued as is shown in figure 10*a*. In this model, where deaths occurred only when a moving individual landed on a Hostile square or left the matrix, the growth curve was very similar to the difference equation analogous to the logistic:

$$N_{t+1} = N_t\{1 + R(1 - N_t/K')\}, \tag{2}$$

in which K' is the upper asymptote and R the intrinsic rate of natural increase.

At no time in the generations following $N_t \sim K'$ were all the Tolerant or Benign areas occupied simultaneously. In other words the total population, N_t, always stabilized at a value less than the carrying capacity of the environment, conventially known as K.

In these primitive simulations it was assumed that hypothetical organisms responded to the density at the geometric centre of population at their time of birth. In later simulations, by dividing the environmental matrix into 156 squares each of 10 × 10 elements, each animal was made to respond only to local density in its own square, or deme, so that, like the population distributions in figure 6, there is a continuum of deme densities. Also, reproduction was introduced, which requires two individuals to find the same coordinates, and reproductive rate was made variable by setting the environmental score −1, 0, 1, 2, 3, etc., where −1 corresponds to Hostile, 0 to Tolerant, and 1, 2, 3,

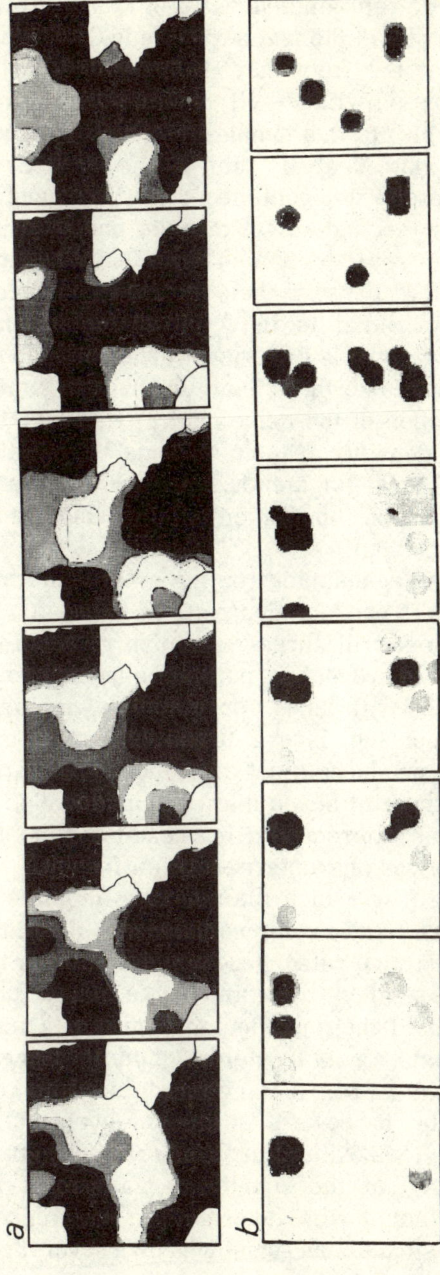

Figure 10 Computer simulation from the Δ-equation of: (*a*) a population, reproducing by binary fission and migrating in response to a central mean density, after rising to a quasi-stable condition (see figure 3*a*); (*b*) a hermaphrodite population becoming chaotic after migration in response to local deme density (see figures 3*b* and 11*b* and *c*). The generations mapped are marked in figure 11*b*.

etc., to Benign with reproductive rates of R, $2R$, $3R$, etc. These values were assigned in the manner already described but with the condition that the frequency distribution of R should be approximately Poisson with $\mu \sim 2R$.

The biology built into this simulation model is more complex and, somewhat, more realistic than in the previous model. In binary fission there is no spontaneous mortality, only death by killing, but in this second set of models, each individual dies immediately after reproduction. The juveniles then migrate, simultaneously, in response to their parents' population density. Movement is now radial to the centre of the deme but, to introduce some reasonable biological variability into this otherwise deterministic stage, there is an uncertainty built into each individual's detection of the deme centre. Having the juveniles respond to the density of their parents' generation is not unreasonable, at least for aphids, although there are endless other ways in which population density may be detected (Wynne-Edwards, 1962).

The same computer-mapping routine used to produce figures 4 and 5 was used to produce figure 10. In figure 10*a* the binary simulation is mapped at three-generation intervals, and like *Spilosoma lutea* (figure 4*a*) it persists almost throughout the geographical range with holes appearing, moving around, and eventually disappearing. Figure 10*b* shows an hermaphrodite simulation which produces a set of islands which grow, move, and decline like those of *Euxoa nigricans* in figure 4*b*.

With a range of different starting conditions, each resulting population curve was characterized by an increase in numbers to an equilibrium from which the numbers fluctuated. Fluctuations were greater in the sexual model than in the binary model. With high reproductive rates some had the appearance of the periodic outbreaks called, by Li and Yorke (1975), 'chaos'. May (1975) has classified the behaviour of equation 2 according to the value of R, that is: globally stable equilibrium when $0 < R < 2$ (figure 11*a*); $2 < R < 2.692$ leading to stable limit cycles with periods increasing in powers of 2; and cycles of aperiodic behaviour when $R > 2.692$. Figure 11*b* shows the trajectory of total numbers (N_t) of the simulation mapped in figure 10*b* compared with that of the deterministic difference equation (figure 11*c*). The simulations sometimes take several generations

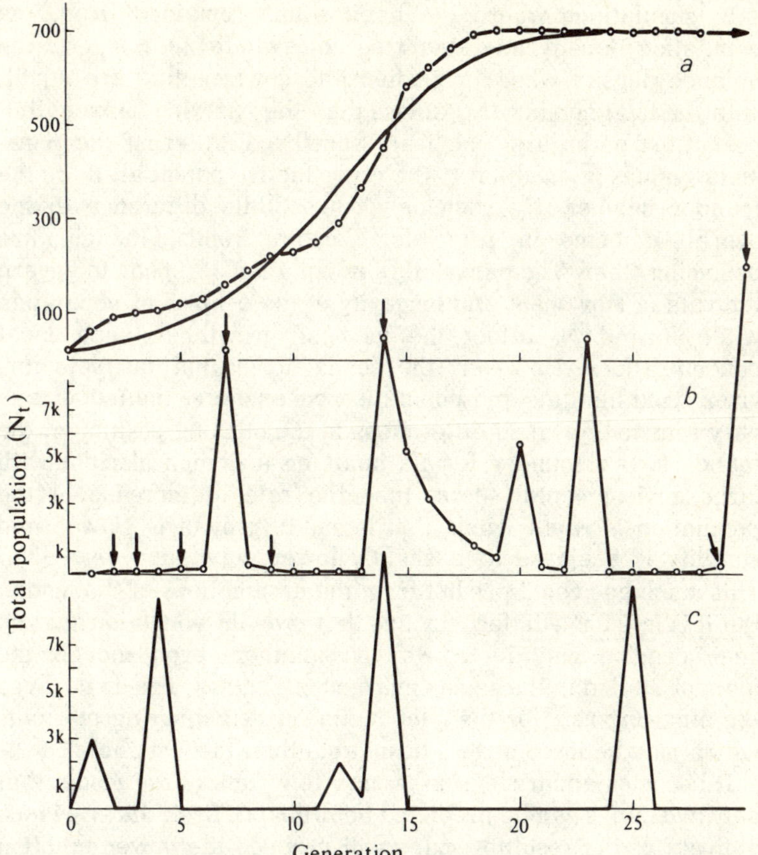

Figure 11 Trajectories in time for the total population (N_t) for:
a Δ-simulation in which the rising population is adequately described by the logistic-like equation 2, drawn in the same figure;
b the Δ-simulation in figure 10*b*, which is chaotic-like,
c the theoretical expectation from equation 2 with $R = 5$ and $N_0/K = 0.02$ (May, 1975).

to return to the endemic level, a feature of natural populations which difference models fail to reproduce because if $N_t > K$, then $N_{t+1} < K$. But with spatial redistribution as well as strong numerical density dependence, the Δ-model permits both rapid and slow returns to K, so there is no 'characteristic return time' (May *et al.*, 1974).

In simulations where the habitat has remained fixed, the population density has stabilized somewhere near to, ρ_0, the balance density where migration and congregation are equal, with the total number (N_t) at less than the carrying capacity but permitting outbreaks which are shortlived. Altering the parameter values in equation 1, the reproductive potential, R, or the founder number, N_0, seemingly makes little difference to the overall outcome in principle. Certain trends, though, are becoming clear. The parameters in equation 1 appear to govern the rate of movement and longevity of the centres of population while R and N_0 affect the rate of growth of such local concentrations. However, the connection is not complete for, given fixed habitats and population parameters, the outcome is very sensitive to small differences in the starting position of the founders. If a founder female lands on a Benign element with large environmental score, then the rate of increase of the population is rapid, while if it is small, growth is slower and stability is achieved at a slightly lower population level (N_t). This much one could predict from the assumptions of the model. But it is less immediately obvious that growth, which is rapid on one occasion may be slower on another, even though the founder settled in the same environmental class. This is because the offspring rely for their fecundity on neighbouring elements which may be more or less hospitable than their mothers'.

It becomes apparent after a very few runs of the model that only two things can be predicted beforehand: first, that variance analysis of the resulting maps will produce the power function required to match natural populations (figure 12) whatever conditions are imposed. Secondly, the population is unlikely to overpopulate its habitat in accordance with common observation of animal populations. A corollary suggests that extinction, of the species as distinct from the population, is a comparatively rare event and will be difficult to cause except by removal, or wide scattering of the Benign environment, a conclusion which most applied biologists would endorse, except when the species is large and the environmental risk is man. In simulations with environmental heterogeneity in time as well as space, it was found that extinction could be forced by reducing fecundity and inhibiting movement simultaneously. In other words, a species occupying a rapidly changing environment must invest in a large

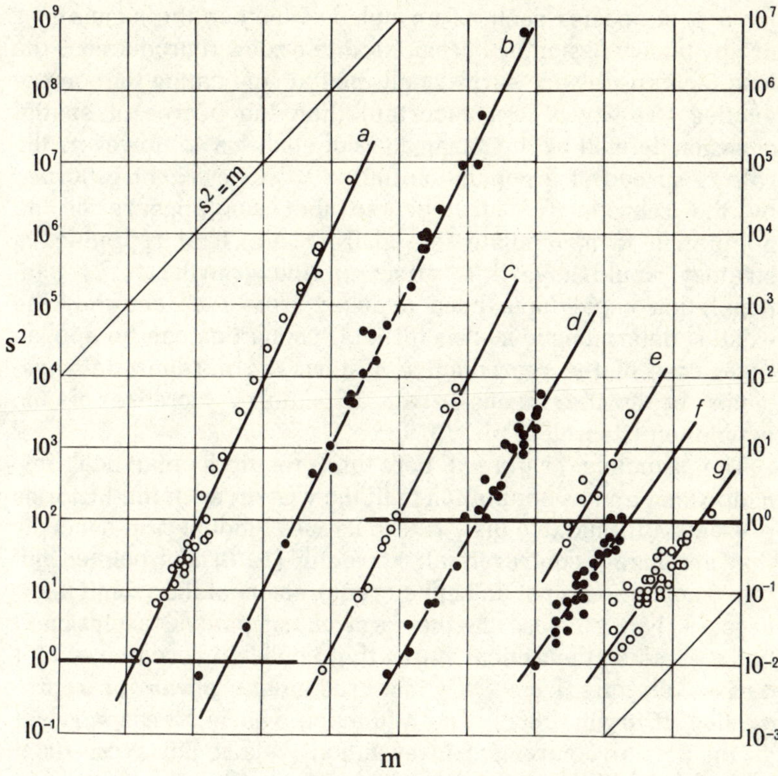

Figure 12 Variance–mean plots for seven Δ-simulations, similar to those for real animals in figure 7. The parameters used in the models and the power function constants are as follows:

	ϵ	ρ_0	p	q	R	b
(a)	3	10	4	3	5	2.19
(b)	3	10	4	3	5	2.06
(c)	33000	1000	1.5	1.4	5	1.87
(d)	33000	1000	1.5	1.4	2	1.70
(e)	0.333	40	4	2	2	1.66
(f)	21000	144	2	1.5	2	1.45
(g)	21000	144	2	1.5	2	1.40

(*a*) and (*b*) are replicates of the hermaphrodite model; (*f*) and (*g*) replicate the binary model; (*c*) and (*d*) compare the effect of raising the reproductive capacity *R* in the primitive model; (*e*) is the result of radial migration in an homogeneous environment (see Taylor and Taylor, 1977).

number of highly mobile offspring, which is almost a definition for a pest species such as an aphid. Whether the simulations are by binary fission or hermaphrodite sexual reproduction, the long-term results are surprisingly similar, indicating that reproductive strategy is less important than the degree of spatial cohesion defined by the parameters of equation 1. However, the rate of spread of a population unit is to some extent governed by the reproductive strategy, so that sex appears as an adaptation to manipulate the environment. That reproductive strategy would have little effect on the growth of the total population could have been deduced from the fact that the existing deterministic models for total population can be applied regardless of the reproductive system, the appropriate parameter being the intrinsic rate of natural increase of the population (Hassell *et al.*, 1976).

The principle presented here still requires analytical formalization, and its adaptation to fit the vast array of life-histories presents a formidable task. It also implies fundamental concepts that are highly controversial. Lewontin (1970) has pointed out that competition is not an implicit requirement of the evolutionary process. Nevertheless, in the form of competitive exclusion it has assumed a theoretical status that is widely accepted, and it derives directly from Darwin's 'redundant power of reproduction' (Darwin, 1868). The Δ-function clearly places survival as the prime requirement in evolution (Wynne-Edwards, 1962) with the avoidance of competition on a par with competition. This has far-reaching implications in ecology and evolution and the model is still new and, as yet, relatively untested. However, the concept of a 'winner' of each conflict being the only effective survivor seems to us to be 'at variance with the facts of nature as we know them'. Otherwise there would have been no end to the evolutionary advantages of total aggression and, since this is evidently not so, there must be a balancing factor which should have its place in the logic of population dynamics and evolution. We suggest that mobility should perhaps be given some consideration.

References

Berry, R. E. and Taylor, L. R. (1968) High altitude migration of aphids in maritime and continental climates. *Journal of Animal Ecology*, **37**, 713–722.

Blackman, R. L. (1974) Life-cycle variation of *Myzus persicae* (Sulz.) (Hom.,

Aphididae) in different parts of the world, in relation to genotype and environment. *Bulletin of Entomological Research*, **63**, 595–607.

Bonnemaison, L. (1951) Contribution à l'étude des facteurs provoquant l'apparition des formes ailees et sexuées chez les Aphidinae. *Annales des Epiphyties*, **2**, 1–380.

Cavalli-Sforza, L. (1962) The distribution of migration distances: models, and applications to genetics. In *Les déplacements humains*, ed. Sutter, J. pp. 139–158. Monaco. Union Européene d'Éditions.

Chiang, H. C. (1960) Effect of the mutilation of wings on the reproduction of the bean aphid *Aphis fabae* Scop. *Entomologia Experimentalis et Applicata*, **3**, 118–120.

Darwin, C. (1868) *The variation of animals and plants under domestication*. Vol. 2. London: John Murray.

Dixon, A. F. G. (1971) The 'interval timer' and photoperoid in the determination of parthenogenetic and sexual morphs in the aphid *Drepanosiphum platanoides*. *Journal of Insect Physiology*, **17**, 251–260.

Dixon, A. F. G. (1976) Reproductive strategies of the alate morphs of the bird cherry-oat aphid *Rhopalosiphum padi*, L. *Journal of Animal Ecology*, **45**, 817–830.

Dry, W. W. and Taylor, L. R. (1970) Light and temperature thresholds for take-off by aphids. *Journal of Animal Ecology*, **39**, 493–504.

Flowers, B. H. and Mendoza, E. (1970) *Properties of Matter*. London: John Wiley.

Forrest, J. M. S. (1970) The effect of maternal and larval experience on morph determination in *Dysaphis devecta*. *Journal of Insect Physiology*, **16**, 2281–2292.

Grinnell, J. (1904) The origin and distribution of the chesnut-backed chickadee. *Auk*, **21**, 364–382.

Hägerstrand, T. (1962) Geographic measurement of migration. In *Les déplacements humains*, ed. Sutter, J. pp. 61–83. Monaco: Union Européene d'Éditions.

Halgren, L. A. and Taylor, L. R. (1968) Factors affecting flight responses of alienicolae of *Aphis fabae* Scop. and *Schizaphis graminum* Rondani (Homoptera: Aphididae). *Journal of Animal Ecology*, **37**, 583–593.

Hamilton, W. D. (1972) Altruism and related phenomena, mainly in social insects. *Annual Review of Ecology and Systematics*, **3**, 193–232.

Hassell, M. P., Lawton, J. H., and May, R. M. (1976) Patterns of dynamical behaviour in single-species populations. *Journal of Animal Ecology*, **45**, 471–486.

Heape, W. (1931) *Emigration, migration and nomadism*. London: Cambridge University Press.

Hutchinson, G. E. (1957) Concluding Remarks. *Symposia on Quantitative Biology*, **22**, 415–427. New York: Cold Spring Harbour Laboratory.

Hutchinson, G. E. (1965) *The ecological theatre and the evolutionary play*. New Haven: Yale University Press.

Ibbotson, A. and Kennedy, J. S. (1951) Aggregation in *Aphis fabae* Scop. 1. Aggregation on plants. *Annals of Applied Biology*, **38**, 65–78.

Johnson, B. (1957) Studies on the degeneration of the flight muscles of alate aphids. I. A comparative study of the occurrence of muscle breakdown in relation to reproduction in several species. *Journal of Insect Physiology*, **1**, 248–256.

Johnson, B. (1958) Factors affecting the locomotor and settling responses of alate aphids. *Animal Behaviour*, **6**, 9–26.

Johnson, B. (1959) Studies on the degeneration of the flight muscles of alate aphids. II. Histology and control of muscle breakdown. *Journal of Insect Physiology*, **3**, 367–377.

Johnson, B. (1965) Wing polymorphism in aphids. II. Interaction between aphids. *Entomologia Experimentalis et Applicata*, **8**, 49–64.

Johnson, B. and Birks, P. R. (1960) Studies on wing polymorphism in aphids. I. The developmental process involved in the production of the different forms. *Entomologia Experimentalis et Applicata*, **3**, 327–339.

Johnson, C. G. (1969) *Migration and dispersal of insects by flight*. London: Methuen.

Kennedy, J. S. (1958) Physiological condition of the host-plant and suspectibility to aphid attack. *Entomologia Experimentalis et Applicata*, **1**, 50–65.

Kennedy, J. S. (1966) The balance between antagonistic induction and depression of flight activity in *Aphis fabae* Scopoli. *Journal of Experimental Biology*, **45**, 215–228.

Kennedy, J. S. (1969) *The relevance of animal behaviour*. London: Imperial College of Science and Technology.

Kennedy, J. S. and Crawley, L. (1967) Spaced-out gregariousness in sycamore aphids *Drepanosiphum platanoides* (Schrank) (Hemiptera, Callaphidae). *Journal of Animal Ecology*, **36**, 147–170.

Lees, A. D. (1960) The role of photoperiod and temperature in the determination of parthenogenetic and sexual forms in the aphid *Megoura viciae* Buckton. II. The operation of the 'interval timer' in young clones. *Journal of Insect Physiology*, **4**, 154–175.

Lees, A. D. (1961) Clonal polymorphism in aphids. *Insect Polymorphism*, 68–79. London: Royal Entomological Society.

Lees, A. D. (1964) The location of the photoperiodic receptors in the aphid *Megoura viciae* Buckton. *Journal of Experimental Biology*, **41**, 119–133.

Lees, A. D. (1967) The production of the apterous and alate forms in the aphid *Megoura Viciae* Buckton, with special reference to the rôle of crowding. *Journal of Insect Physiology*, **13**, 289–318.

Lees, A. D. (1975) Aphid polymorphism and 'Darwin's demon'. *Proceedings of the Royal Entomological Society of London*, **39**, 59–64.

Lewontin, R. C. (1970) The units of selection. *Annual Review of Ecology and Systematics*, **1**, 1–16.

Li, T.-Y. and Yorke, J. A. (1975) Period three implies chaos. *American Mathematical Monthly*, **82**, 985–992.

Lidicker, W. Z., Jr. (1962) Emigration as a possible mechanism permitting the regulation of population density below carrying capacity. *American Naturalist*, **96**, 29–33.

Lowe, H. J. B. and Taylor, L. R. (1964) Population parameters, wing production and behaviour in red and green *Acyrthosiphon pisum* (Harris) (Homoptera: Aphididae). *Entomologia Experimentalis et Applicata*, **7**, 287–295.

MacArthur, R. H. (1972) *Geographical Ecology*. New York: Harper and Row.

McGuire, J. U., Brindley, T. A., and Bancroft, T. A. (1957) The distribution of European corn borer larvae *Pyraustia nubilalis* (Hbn.) in field corn. *Biometrics*, **13**, 65–78.

May, R. M. (1975) Biological populations obeying difference equations: stable points, stable cycles, and chaos. *Journal of Theoretical Biology*, **51**, 511–524.

May, R. M., Conway, G. R. Hassell, M. P., and Southwood, T. R. E. (1974) Time delays, density-dependence and single species oscillations. *Journal of Animal Ecology*, **43**, 747–770.

Moericke, V. (1950) Über das Farbsehen der Pfirschblattlaus (*Myzodes persicae* Sulz.). *Zeitschrift für Tierpsychologie*, **7**, 265–274.

Preston, F. W. (1969) Diversity and stability in the biological world. In *Diversity and Stability in Ecological Systems*, pp. 1–12. Brookhaven Symposium in Biology, 22. Upton, New York: Brookhaven National Laboratory.

Rose, D. J. W. (1972) Times and sizes of dispersal flights by *Cicadulina* species (Homoptera: Cicadellidae), vectors of maize streak disease. *Journal of Animal Ecology*, **41**, 495–506.

Roughgarden, J. (1974) Population dynamics in a spatially varying environment: How population size 'tracks' spatial variation in carrying capacity. *American Naturalist*, **108**, 649–664.

Shaw, M. J. P. (1970a) Effects of population density on alienicolae of *Aphis fabae* Scop. I. The effect of crowding on the production of alatae in the laboratory. *Annals of Applied Biology*, **65**, 191–196.

Shaw, M. J. P. (1970b) Effects of population density on alienicolae of *Aphis fabae* Scop. II. The effects of crowding on the expression of migratory urge among alatae in the laboratory. *Annals of Applied Biology*, **65**, 197–203.

Shaw, M. J. P. (1970c) Effects of population density on alienicolae of *Aphis fabae* Scop. III. The effect of isolation on the development of form and behaviour of alatae in a laboratory clone. *Annals of Applied Biology*, **65**, 205–212.

Skellam, J. G. (1951) Random dispersal in theoretical populations. *Biometrika*, **38**, 196–218.

Southwood, T. R. E. (1962) Migration of terrestrial arthopods in relation to habitat. *Biological Reviews of the Cambridge Philosophical Society*, **37**, 171–214.

Taylor, L. R. (1957) Temperature relations of teneral development and behaviour in *Aphis fabae* Scop. *Journal of Experimental Biology*, **34**, 189–208.

Taylor, L. R. (1959) Abortive feeding behaviour in a black aphid of the *Aphis fabae* group. *Entomologia Experimentalis et Applicata*, **2**, 143–153.

Taylor, L. R. (1961) Aggregation, variance and the mean. *Nature, London*, **189**, 732–735.

Taylor, L. R. (1965) Flight behaviour and aphid migration. *Proceedings of the North Central Branch of the American Association of Economic Entomologists*, **20**, 9–19.

Taylor, L. R. (1970) Aggregation and the transformation of counts of *Aphis fabae* Scop. *Annals of Applied Biology*, **65**, 181–189.

Taylor, L. R. (1971) Aggregation as a species characteristic. In *Statistical Ecology*, ed. Patil, G. P., Pielou, E. C., and Waters, W. E., vol. 1, 357–377. University Park, Pennsylvania State University.

Taylor, L. R. (1974) Monitoring change in the distribution and abundance of insects. *Report of the Rothamsted Experimental Station for 1973*, Part 2, 202–239.

Taylor, L. R. (1975) Longevity, fecundity and size; control of reproductive potential in a polymorphic migrant, *Aphis fabae* Scop. *Journal of Animal Ecology*, **44**, 135–163.

Taylor, L. R. (1977) Migration and the spatial dynamics of an aphid, *Myzus persicae*. *Journal of Animal Ecology*. **44**, 411–423.

Taylor, L. R. and Taylor, R. A. J. (1977) Aggregation, migration and population mechanics. *Nature, London*, **265**, 415–421.

Taylor, L. R. and Woiwod, I. P. (1978) The density-dependence of spatial behaviour and the rarity of randomness. *Journal of Animal Ecology*, **47** (in press).

Taylor, R. A. J. (1978) The relation between density and distance of dispersing insects. *Ecological Entomology*, **3** (in press).

Wadley, F. M. (1931) Ecology of *Toxoptera graminum*, especially as to factors affecting importance in the Northern United States. *Annals of the Entomological Sociey of America*, **24**, 325–395.

West Eberhard, M. J. (1975) The evolution of social behaviour by kin selection. *Quarterly Review of Biology*, **50**, 1–33.

Wilson, E. O. (1975) *Sociobiology: The New Synthesis*. Cambridge, Mass.: Harvard University Press/Belknap.

Wynne-Edwards, V. C. (1962) *Animal Dispersion in Relation to Social Behaviour*. Edinburgh and London: Oliver and Boyd.

Spacing mechanisms and adaptive behaviour of Australian Aborigines

JOSEPH B. BIRDSELL

Department of Anthropology, The University of California, Los Angeles, USA

Introduction

This paper is a preliminary formulation of ongoing research on the human ecology of the Australian Aborigines. The early exploratory phases of the endeavour have long been published (Birdsell, 1953). A complex of environmental variables has been included in the recent analysis and these yield very high correlations with the dependent variable, area of the tribal domain. It is now clear that these Aborigines achieved and maintained an effective homeostatic living system.

The natives of Australia are an ideal population for this type of study. Firstly, these hunters occupied an entire continent with no interference from peoples with higher types of economy, prior to the appearance of the first colonial Europeans. Secondly, in spite of a great diversity of environments, ranging from the harshest of deserts to tropical rain forest, the approximately 600 tribes which constituted the population were remarkably uniform in their technology. Finally, they are certainly the best studied of all living, hunting and gathering peoples, and so more is known of the factors which structure their populations. Special notice should be taken of the major contributions made by Norman B. Tindale who has devoted more than a half century to fieldwork among the Aborigines. This writer has had the good fortune to have collaborated with Tindale both in the field and in numerous intellectual ventures over nearly 40 years. This is a debt which I gratefully acknowledge.

The aboriginal Australians stand as a special example of the operation of homeostatic processes among economically simple peoples. But other generalized hunting and collecting people, even if less well known, give evidence of the same kind of stable population equilibrium. The Great Basin Shoshoni reveal a generally similar pattern as indicated by Vorkapich (1976). The same is true for the very limited sample of hunting peoples available from Africa, as explored by Martin and Reed (1976). There is good reason to believe (Birdsell, 1968) that the same general type of homeostatic system existed far back into time, among the generalized hunters of the Pleistocene. This projection is based upon the components of behaviour found among the Australians which can reasonably be attributed to earlier peoples living at the same simple economic level.

The triangle of basic ecological relationships

In the next section it will be shown that tight quantitative relationships exist between a complex of environmental variables and aboriginal density in Australia. This correlation should not be interpreted to indicate the operation of environmental determinism as the sole factor. Proper weight must be given to the role of population biology on the one hand, and of social organization on the other, in evaluating the equilibrium system of these people.

Figure 1 shows a diagrammatic triangle involving three polar relationships, each of which contributes to the overall living system. The apices of this triangle are abstracted as separate poles. In fact they are not completely isolated, but each interacts with the other adjacent ones. The environmental apex includes all of the variables in the overall environment, as well as the other living organisms contained within it. Biology refers to the life processes of the population, including its reproductive practices, all of its demographic attributes, and whatever factors may influence life and death within it. The social pole covers the population aspects of social structure, and all forms of social and cultural behaviour. Each of these polar categories interacts in complex fashion with both of the others.

This type of formulation stands in marked contrast to those offered by earlier generations of anthropologists. In America, both Franz Boas and Alfred Kroeber separated the poles of biology and environment from that involving social or cultural behaviour. Both of these great pioneers seemed to feel that

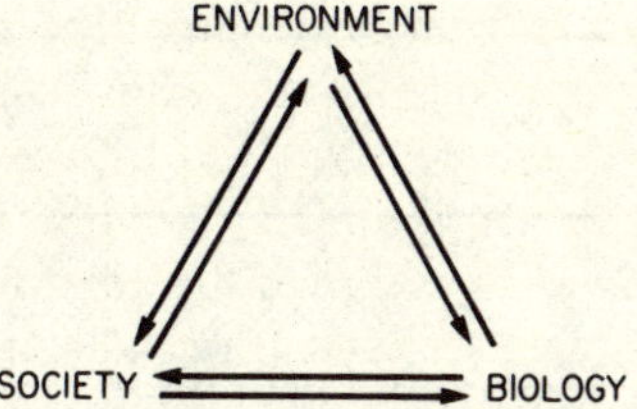

Figure 1 Model of tripolar relationships in human ecology.

population behaviour stood above and separate from them. Kroeber even formulated his position into his famous dichotomy of the organic and the superorganic. In more recent decades the research of human biologists and archaeologists has demonstrated the fallacy of considering social organization in a vacuum with no feedback from the environment or populational biology.

Ecological analysis of aboriginal densities in Australia

An early study (Birdsell, 1953) determined that a simple correlation coefficient of 0.81 existed between the area of 123 basic Australian tribes and mean annual rainfall. The form of the relationship in terms of density is shown in figure 2. This basic series included all tribes for which boundary information was then available (Tindale, 1940), and whose total water resources were derived from rainfall within their own boundaries. This restriction excluded all coastal tribes and all riverine tribes below the headwaters level. As an additional precaution a belt of tribes lying immediately to the west of the easterly boundaries of the initiation rites of circumcision and subincision was also eliminated on the basis that tribal fragmentation had followed in the wake of the diffusion of these ceremonies. Subsequent analysis (Birdsell, 1973) justified this decision.

An important step in this early analysis was the projecting from tribal area to the size of tribal populations. Kryzwicki (1934) concluded that, on the average, Australian tribes consisted of approximately 500 persons. His data were in many ways erroneous, but self-compensating, so his figures still stand

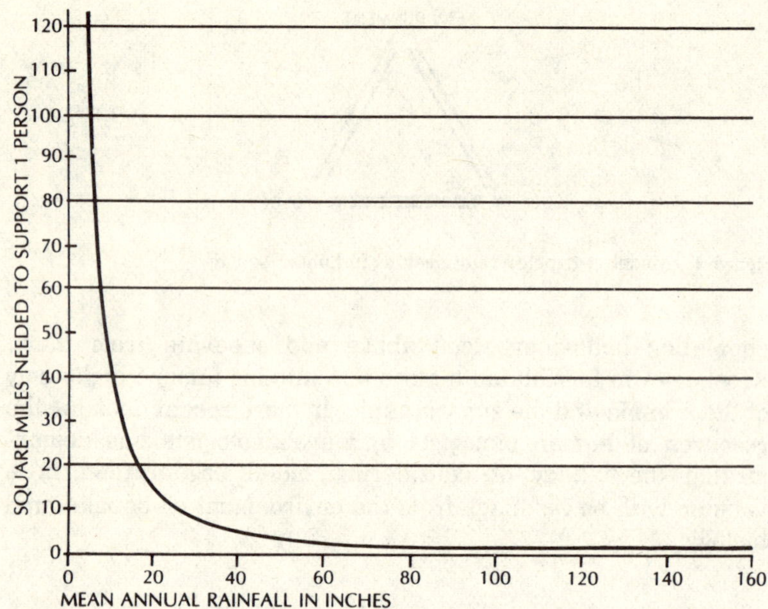

Figure 2 The form of the relationship between mean annual rainfall and aboriginal density (1953 data).

as a fair estimate. In the 1953 analysis, 45 estimates of the size of tribal populations were utilized as a check against the estimate of population size derived from the actual tribal area judged against its area predicted by the rainfall equation. These data were neither sensitive nor sufficient, but they tended to substantiate the estimate by Kryzwicki. Finally, the correlation coefficient of 0.81 yielded an unexplained variance of only 35 per cent in the system. This was judged to indicate a central numerical tendency for the Australian tribal population to be stabilized around 500 persons. Since computers were not available at that date, the analysis rested there.

Current research project

An opportunity arose in 1973 to become a Research Scholar at the Australian National University in Canberra, Australia. During the year spent there in residence, raw data were collected for as many Australian environmental variables as had then been published. Norman B. Tindale, also then a Research Scholar at ANU, made available to me all of his raw data in the tribal monograph he was then revising. This magnificent contribution to human territoriality, which delineates tribal boundaries throughout the Australian continent, has since been published (1974).

Of Tindale's nearly 600 tribes, a total of 183 were chosen as suitable for this analysis. The same ecological restraints were used as in the earlier paper (1953), so that only those tribes depending upon rainfall falling within their boundaries were utilized. All coastal tribes were excluded, as well as those below the headwaters of all rivers, ephemeral or permanent. The fragmentation belt to the west of the easterly diffusing rites of subincision and circumcision was again omitted. But based upon the analysis of 1973, this time it was decided to include a few tribes lying at the northern end of these boundaries where fragmentation was not evident.

Ecological variables utilized

A total of 66 primary environmental variables was used in the computer analysis. In addition 19 sorting variables were available to regroup and refine the basic data. The dependent variable consisted of the tribal area as identified, outlined, and measured by Norman B. Tindale (1974).

The primary variables group themselves into six clusters.

1. Of these the most important was a series of 17 factors measuring rainfall or related to it. Examples would include median annual rainfall, the number of rain days, and Martonne's duration of the arid period, which is a negative expression of rainfall.
2. A second cluster of primary variables related to temperature and numbering 14 in all. Among these were mean annual temperature, duration of maximum heat wave, and the number of days of frost.
3. A third category included four variables representing simple measures of solar radiation. They occurred in pairs, involving total radiation for both January and July, and net radiation for the same seasons of the year.
4. A series of four other simple and independent variables

included soil fertility and the percentage of tribal domain in sand hills in desert areas among others.

5. Fortunately, a series of complex botanical variables had been modelled for the entirety of Australia by botanist Ray Specht. These were made available to me through his courtesy prior to publication. As examples may be mentioned annual net photosynthetic index with foliage projective cover varying seasonally, and total growth index for climax vegetation including both tops and roots.

6. Finally, a sixth set of complex variables, numbering 24 in all, included such factors as Thornwaite's evapotranspiration, Gentilli's annual phytohydroxeric index, and Moore's summer moisture index.

Most of these primary variables were quantitatively derived from instrument readings. In the best cases the years of record were more than a half a century in length, but others covered a considerably shorter time span. No selective judgement was used in rejecting available data on the grounds that simplistic personal judgement could not evaluate their contribution in a multivariate analysis of considerable complexity. As might have been anticipated, seemingly obscure or unpromising variables did influence aboriginal man's occupancy of the continent. For example, the standard deviation of annual evaporation, and temperature lag in days behind radiation, both unlikely sounding variables, each made their contributions.

Preliminary results of this analysis

Median annual rainfall alone gave a Pearsonian coefficient of correlation of 0.794 against the size of tribal areas. This value is somewhat lower than the 0.81 obtained for a smaller series of tribes in 1953. This coefficient leaves about 37% of the variance in the system unexplained. Yet it can be read as indicating that the overall environment exerts powerful determinism upon the sizes of the tribal areas occupied in Australia, and further that the numbers of individuals comprising the dialectal tribe approach a constant value. This has previously been estimated as close to or perhaps slightly below 500 persons.

One of the primary goals of this research was to investigate the degree to which the environment in fact imposed restraints upon human occupancy in aboriginal Australia. Further exploration of the data made it evident that the summer rainfall regime, coming in with the monsoon in the north, provided higher degrees of environmental determinism than did the winter rainfall pattern in the southern half of the continent. Using five per cent increments, it is found that the correlation coefficient for median annual rainfall maximized for those tribes in which summer rainfall equalled or exceeded 45 per cent of the total recorded rainfall. This cut-off point yielded an optimum series of 86 tribes which extended across the northern portion of the continent as shown in figure 3. The bivariate correlation between median annual rainfall and tribal area reached 0.869 for this series, which indicates an unexplained variance of less than 25 per cent in the system tested. Proceeding with this series to a multiple step regression analysis, a total of 66 environmental variables yielded a raw coefficient of 0.989.

This value for the multiple step coefficient of correlation is inflated, so it seems wise to look briefly at several areas which

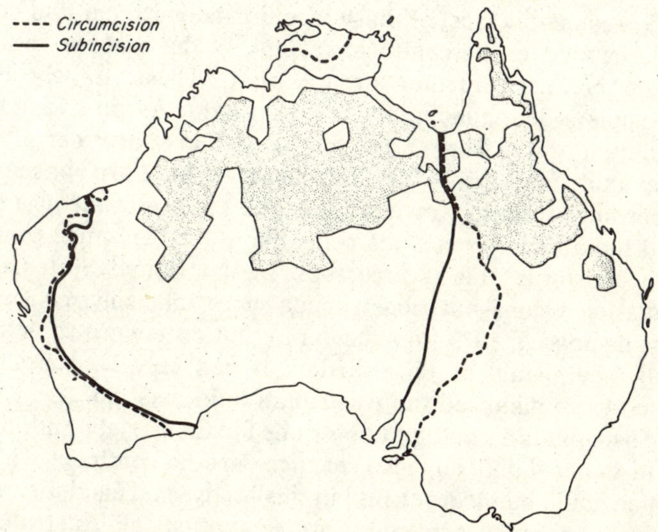

Figure 3 The distribution of 86 tribes with 45 per cent or more winter rainfall in the 1976 series.

may contribute or detract from its face value. The first of these is statistical in nature. It involves the fact that the environmental complex used contains many factors related to each other, indeed showing a high degree of correlation with one another. This produces a situation known as multicollinearity. It exists in extreme form when intercorrelations within the range of 0.8 to 1.0 occur between the independent variables. Certainly a considerable degree of multicollinearity exists in this analysis, and testing to discover serious distortion in the correlation values will await further analysis. A second statistical point involves the fact that no lower limits of the F ratio for variable inclusion were set. With conservative treatment, including a high F for inclusion, R^2 adjustment and Mallow's *Cp* calculated, the coefficient for the multiple step regression should reach 0.90 to 0.91. For these values the unexplained variances fall to 19% and 17%, and are figures of proper magnitude.

There are an opposing category of factors which would tend to diminish the proper values for these coefficients of correlation. They involve a variety of errors in the ascertainment of the data. Even in the earlier study (1953) six categories of errors were identified in the data used. No corrections could be made and presumably most of these would randomize out and so have little impact on the coefficient value. In the present analysis a dozen or more further errors exist. These largely involve incomplete or inadequately reported data. As an example, the radiation data from Australia used here are certainly inadequate, and probably inaccurate. They are based upon recordings from too few stations, and the use of the assumption that the albedo, or surface reflectivity, for the entire continent, is a constant. This is incorrect. In the United States, where radiation values have been much more intensely measured, it was decided in 1972 that results to that date were so erroneous that they should be disregarded. It can only be guessed how much more distorted the Australian data must be.

Australia is a continent notorious for the variety and intensity of its mineral deficiencies. Yet there are no continent-wide data which could be incorporated in this analysis. This deficit results from a natural economic preoccupation of Australian researchers with the more productive portions of their continent. How much a full knowledge of mineral deficiencies on the continent would contribute to the present study is of course unknown. Since the relation between aboriginal man and his land in Australia is highly deterministic, the writer suspects that these, and other errors, and omissions in data, are sufficient to partially compensate for the statistical hazards of multicollinearity.

The other ecological poles

That environmental determinism exerts strong influence upon the Australian Aborigines is hardly to be doubted. But the feedback relationship between environment, population biology, and society depicted in figure 1 suggests that complex factors produce the homeostatic system evident among these people. Certainly the occasional exercise of Liebig's law during very bad seasons must be granted. But on a year-to-year basis, the aboriginal population must have maintained a roughly optimum level. The evaluation of this level numerically is difficult. Some suggestions from experiments in nature (Birdsell, 1957) suggest that perhaps a comfortable population level might be in the range of 50 to 60 per cent of the maximum carrying capacity in a given environment. Homeostasis at this level could be maintained by the complex of forces previously invoked.

Population biology

Population biology covers a wide range of phenomena. But the maintenance of a balanced living system primarily requires a careful control of reproduction. Stable systems imply stable numbers. Every successful population has inbuilt excess fertility and this must be controlled among complex bisexual animals, including economically simple men.

The intrinsic rate of increase can be defined, after Birch (1948), as the number of times that a population multiplies per generation in an unlimited environment. For man, three isolated island communities provide empirical evidence suggesting an intrinsic rate of increase involving a doubling per generation (Birdsell, 1957). Data for hunters and gatherers are harder to obtain. Therefore the important natural experiment reported early on by Cudmore (1893) is particularly appropriate since it involves aboriginal Australians.

This experiment involved the so-called Nanja horde of the Danggali tribe in the Darling River section of western New South Wales. The original account merely indicated that the young man, Nanja, had gone bush with one or two young women and remained out of contact with colonial Whites for 30 years. When rediscovered this little group totalled 29 persons and involved three generations. Norman B. Tindale, who has reworked all of the available materials (personal communication) indicates that each of the two original women bore ten viable and surviving children. Although the little group lived in desolate country and were dependent on water from plant roots, they prospered to the extent that their numbers increased five-fold in the first generation. This is eloquent testimony to the need for controlling human fertility if population stability is to be maintained.

Infanticide

Australians had several options for the throttling of human fertility, but only one seemed to have been exercised. Sexual abstinence is so rare as to be seldom mentioned in the literature and not likely to have been widely practised. Abortion had been referred to by a few early writers. They suggest the process involved a pummelling of the abdomen of a pregnant woman to produce miscarriage. Infanticide was a widespread practice and commonly referred to by early explorers and colonists, as well as by later anthropologists. Infanticide in the eyes of European law constitutes murder, so that in post-contact time there has been a natural reticence among the natives concerning it.

Aberrant sex ratios provide one indirect measure of the effects of infanticide, if it is practised preferentially on one sex or the other. There is general agreement that female infants were preferentially killed in aboriginal Australia. Therefore data showing a sex ratio biased in favour of males is evidence for this particular practice.

The following data are based upon genealogies collected during 1953 and 1954 by Norman B. Tindale and the writer while working around the fringes of the Western Desert of Western

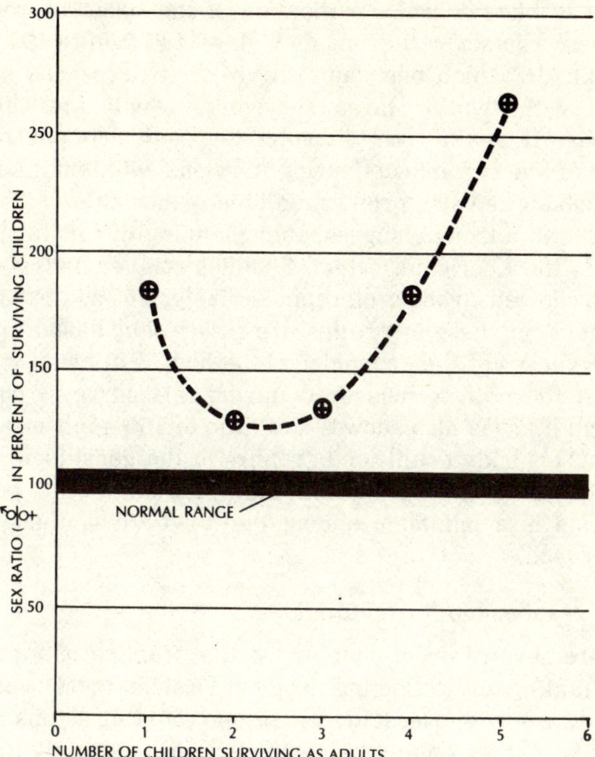

Figure 4 Sex ratios of surviving children in 194 precontact families.

Australia. They involve 194 matings judged by context to have been precontact at inception and during the woman's child-bearing years. The number of children remembered as surviving toward adulthood are plotted in figure 4 against the sex ratio, indicating the percentage of males to females. A striking relationship emerges with a large bias in the male direction in families of all sizes from one to five children. The few families with six and seven children were too rare for inclusion.

Other genealogical data, derived from families with large numbers of surviving children in situations such as on missions where infanticide would be impractical, indicate that the sex ratio at birth among these natives approximates 100, and so is biologically normal. Therefore these aberrant sex ratios point strongly to behavioural modification of the basic reproductive processes. The sex ratio runs as high as 186 men to 100 women in families in which one child survives. It decreases in those families with two or three survivors, which are the most numerous. It again rises sharply, to reach 260 per cent, in families of five surviving offspring. Obviously systematic infanticide involving females preferentially is responsible.

These data strongly suggest that planning of family size was built into the aboriginal sytem. Families smaller or larger than average showed strong preference for males among the surviving children. That these sex ratios are no artifact based upon the inherent biases of the genealogical method, is attested to by the fact that the natives measured during this survey from these same tribal areas also show a sex ratio of 150 adult men to 100 women. This body count lends reality to the genealogical data. When the data are cumulated they suggest the 15 per cent infanticide is a minimum among the newborn of these desert tribal groups.

Factors responsible for infanticide

There are several overt motives for the practice of infanticide among hunting and collecting peoples. First, in these societies a woman can only successfully nurse one child at a time. Diet is such that weaning cannot be completed until certainly the third year, and frequently not until after the fourth year, following birth. If the renewal of ovulation was under complete hormonal control in the human species, this would pose no problem. But there is ample evidence that in at least some women, and so possibly in all populations, new pregnancies can follow early on and before the last child is fully weaned. Thus a need for spacing children arises.

A second pressing factor involves maternal mobility. Even where hunters need not shift camp every day, they still will do so frequently. Further, the women fulfill their economic obligations by collecting and gathering food each day and, indeed, contribute something like 70 per cent to the total family diet. While a woman can conceivably carry one child on an eight mile foraging trip, she obviously cannot carry two. In addition to her infant, she has digging sticks, wooden dishes, and other equipment to transport. So until a child is able to accompany its mother on foot any extra children could seriously handicap her economic activities. Films of precontact Aborigines in the Western Desert show children as old as 5, 6, and possibly 7 years being carried crosswise on the small of their mother's back on journeys in which camp was moved.

A third factor of supreme importance in the maintenance of a homeostatic living system involves the fact that women are the funnels through which fertility is poured. The control of their numbers is the first step in population planning.

There is a fourth factor which helps explain the preference for saving male children. Aboriginal society in Australia is male oriented. It is based upon landowning patrilocal groups, and a culture in which male values are strongly emphasized. Not only is a son a potential defender of the land rights of his parents, but he is the provider of highly esteemed animal foods. Sharing is so finely regulated that parents obtain the choice portions of a son's kill, and so naturally sons are held in high esteem.

Even among anthropologists good data on the practice of infanticide are all too rare. I know of none in which population control is implied or recognized except in the following account. Dr Joan Greenway, fortunately a most perceptive observer, has provided the following personal communication which I feel privileged to quote:

> 'Infanticide among the Wanindilyaugwa of Groote Eylandt was controlled by the "boss grannies"—senior women whose dominant status was achieved by age, wisdom, and personality factors rather than by particular kinship affiliation. The decision to kill an expected child is made before its birth, and death was accomplished by suffocation. A boss granny's hand covered the infant's face as soon as it appeared, so that the child did not ever breathe. This meant, of course, that in the aboriginal view the baby was not killed, but simply prevented from living.'

The implications of this statement extend beyond its obviously rich contents. Previous observers have reported that a mother about to give birth is accompanied into the bush by a 'granny woman', but none have specified her role. This quotation makes clear that the granny woman was there to express the opinion of a small society, the local group, as to whether the new birth could be accepted as a living recruit, or whether population circumstances precluded it. She was present then as an agent of population regulation. Certainly the mother did not need outside opinion to know whether the new infant would present an insurmountable nursing problem, or could not be transported in her daily round of duties. It may seem surprising that population control this sophisticated would be found in a people as apparently simple as these Aborigines.

There remains a problem of whether the 15 per cent rate of infanticide estimated from the 194 precontact families previously examined is a correct estimate, or represents a minimum one. Those data did not include the killing of infants, both male and female, who were simply murdered to meet the spacing requirements of the mother. They also excluded children simply done away with because they could not at the time be recruited into the small local society without undue population increase. One of the few early estimates of the overall frequency of infanticide was made by Mr S. Gason (Curr, 1886) who was stationed for six years in very early days as an officer of the South Australian Mounted Police in the country of the Dieri tribe. It is his opinion that these people killed as many as 30 per cent of the infants born. This estimate would not conflict with the completed figures based upon the extension of the values derived from the genealogical data considered earlier. Additionally, neonatal mortality must have been relatively high in precontact times. But there are no data upon which estimates of rates can be based in Australia. No doubt the actual practice of infanticide took into consideration neonatal losses to the society.

Social organization

Anthropologists customarily consider a great variety of social groups under the heading of social organization. Here our concern is with but three basic ones, the family, the local group or band, and the dialectal tribe. It is appropriate to ignore age groupings, sex groupings, and for the time being totemic clans. Nor is there any need to become involved with the complexities of Australian kinship at this point, nor the various types of sections into which society may be divided to regularize marriage, nor yet the broad array of classificatory relationships.

The family in aboriginal society

The family is the basic unit in Australian aboriginal society. Its members include an adult male, his wife, or occasionally wives, and their children. The offspring may have been sired by the man, in which case the family is a biological unit. Insofar as some children may not be by him, it is still a social family. Polygyny is permitted throughout the continent, but only a few fortunate men attain this marital state. Among Western Desert tribes 85 per cent of the marriages are monogamous, while among the remaining 15 per cent, a man may rarely have up to five wives. In north coastal Australia 'big men' are recorded as having as many as twenty wives, but these are not present in a family all at one time, because of the cycle of marriage and death. Consequently many more men remain unmarried there.

Living adults show a sex ratio of 150 males per 100 females, so that polygyny would seem difficult to achieve. But in general men do not receive their wives until into their third decade of life, whereas the girls are married at or before puberty. These age differences provide differential numbers in the population pyramid, and enough young girls to keep the system working. In general the old men obtain young girls as wives, whereas young men frequently inherit a brother's widow as their first and possibly only woman.

Precontact family attributes

The number of adult offspring surviving per family is shown in figure 5. These data are based on the same 194 families occurring in precontact times taken from genealogies obtained from tribes in the Western Desert. The figures have previously been used in the discussion of infanticide. This practice, in concert with population control, maximizes the number of families in which two or three children grow up. None were recorded with eight surviving offspring, and only 2 per cent had as many as seven surviving children. It is probable that there were more families than the 3 per cent shown in which no children survived. This follows from the artifact in the genealogical method in which families with no survivors would tend to be disproportionately forgotten in the recounting of relationships.

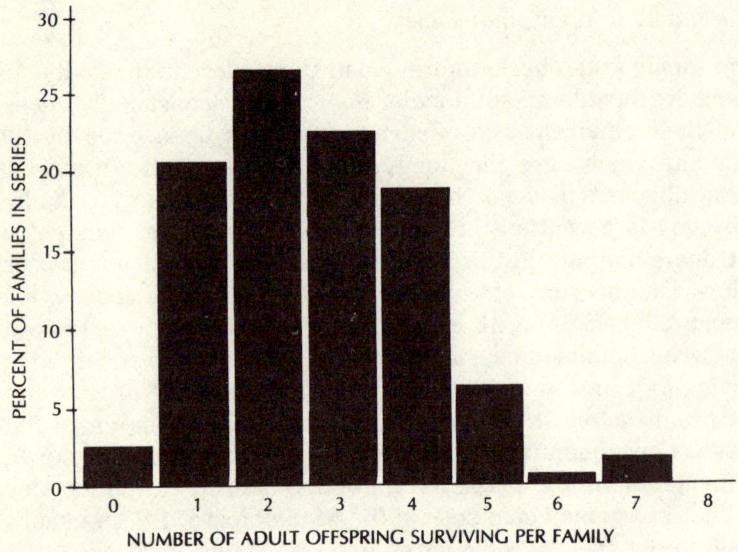

Figure 5 Frequency of surviving offspring in 194 precontact families.

The frequencies show a truncated Poisson distribution. The mean number of children per family in these data is 2.66, somewhat in excess of the 2.00 expected in a population in stable equilibrium. The discrepancy arises from the fact that informant memories do not specify children who exactly reach the point of middle reproductive years. By this criterion, undoubtedly, a number of younger individuals are remembered. The standard deviation for surviving children is 1.43, and the variance is 2.04.

The local group or band

The local group is the largest social unit which functions on a regular face-to-face basis. It owns the hunting and gathering rights of its country. Members of other groups can enter and hunt over this land only upon invitation. The adult members of the band consist of men who are born there and the women from other local groups who marry them. It is strictly exogamous and no marriages are tolerated between members. Daughters born into the local group marry out and so are lost to other bands. In a very real sense the local group can be considered as an extended patrilineal family exercising its territorial rights to land which has long been in their male lineage. The men of a band feel a very spiritual attachment to their country and this is one of the stabilizing influences in aboriginal society. They are reluctant to leave it for long and if forced to do so, as by drought, return at the first opportunity. Their religious ties with their land are so strong as to compel them to try to return to it to die if possible.

By its nature the local group consists of a series of families closely related in the male line. Since the processes of birth and death contain stochastic factors, it might be expected that the local group size would fluctuate and vary considerably. This does not seem to be the case generally. Data provided by N. B. Tindale (personal communication) indicates that for 11 bands seen by him under precontact conditions in desert country, the numbers averaged 25 persons of both sexes and all ages. The range about this mean value was small, and it can be inferred that in the less well endowed regions of Australia band size is in fact under rather tight controls. It has been pointed out elsewhere (Birdsell, 1968) that local group size of this same magnitude characterizes the Kung Bushmen of South Africa, and the Bihor, a hunting and gathering group in North India. Early observers' accounts throughout much of the interior of Australia indicate local group size of about the same magnitude.

But food resources are much richer in the coastal regions of northern Australia. There local group sizes increase seasonally to reach 50 or even 100 individuals. Where local and band resources are abundant, local groups of this size do not seriously handicap the efficiency of hunting and gathering. These concentrations could exploit marine life without loss of efficiency, and did so. It almost seems that there existed a very reasonable aboriginal drive to maximize numbers, and so social interactions, in these local groups, to the extent that food resources allowed.

Band geometry and tribal dialects

The effect of variations of local group geometry on the distribution of linguistic differences was explored in a model (Birdsell,

1958). It succeeded in identifying some of the factors contributing to the Australian situation. With all face-to-face interactions on the continent limited by mobility on foot, distance operated as an isolating factor between individuals, bands, and tribes. If distance alone acted as an inhibitor to communications, then dialect differences should proceed smoothly and continuously from one end of the continent to the other. But this in fact did not happen. Tribes are each characterized by their own dialect, and so the clumping of communications' patterns were characteristic in aboriginal society.

The model utilized a series of contiguous hexagons, each representing a local group territory. This geometry was preferred over that of square land holdings since the tribal configurations themselves tend toward a six-sided configuration. It was assumed that the local group teritories contained within the tribe would approximate a similar geometry. A series of model variations was examined. A tribe was initially assumed to consist of a single band, then six hexagonal bands, and finally in the largest version of 19 six-sided local groups.

It was assumed that local group amity and enmity cancelled out in the long run, and so an equal density of communications crossed each joint boundary. The models showed that as the number of local groups increased within a tribe, the proportion of contacts across tribal boundaries with the local groups of other tribes markedly diminished. An increase in the number of contained bands within the tribe internalized the pattern of communications. But the ideal predictions of the model were not met with in aboriginal reality. This density of intertribal communications cannot be evaluated in precontact Australia, but it is assumed to correlate directly with the ratio of intertribal marriages. With a graph derived from variant models it could be shown that if a tribe contained 20 local groups, 40 per cent of the communications could be predicted as extratribal. Presumably a similar proportion of intertribal marriages would also be contracted. Tindale (1953) has demonstrated that a generalized figure for intertribal marriages in Australia is somewhat less than 15 per cent instead of the 40 per cent predicted. This result clearly indicates inhibiting factors across tribal boundaries so that marital behaviour fell well below model expectation. The deficit in intertribal marriages is substantial and, based upon the above figures, amounts to more than 60 per cent. Thus the territorial properties of the local groups, combined with their geometry, tend to produce a centripetal social and linguistic configuration. These processes tend to stabilize the linguistic pattern and social behaviour within the dialectal tribe.

The nature of the Australian tribe

The concept of the tribe has been used by anthropologists so ambiguously that it can apply to any geographical aggregation of people who are less than a nation and more than a local group. But the tribe in aboriginal Australia has a specific set of attributes, although unfortunately there is no complete consensus among cultural anthropologists as to what these constitute. First to be noted is the fact that tribes on this continent have no political organization. Nor is there any identifiable leadership aside from the older men who constitute an informal gerentocracy and are important in all decision making. Primarily the tribe is a language unit and hence can be called a dialectal tribe. Language is not completely homogeneous within such tribes, but tends to be. Even the fact that linguists, in their recent fine-grained analyses, find some exceptions to this, the observation remains generally true.

Beyond dispute is the fact that a tribe is a collection of local groups geographically adjacent to each other. The tribal territory is the sum of such contiguous bands. Further, tribes are known by distinct names, both by themselves and by others. Proximity in space and in descent produces a complex of customs and laws in which a tribe generally differs a little from adjacent ones. But the best definition of a tribe involves the reproductive behaviour of its members. Since all local groups are exogamous, and the rate of marriage intertribally is low, *the tribe is best defined as a minimal breeding population.* Social anthropologists will resent this intrusion of biology into their area, but it is a rigorous definition which has no exceptions in Australia. It has a certain great advantage in that the Aborigines themselves recognize the members of their tribe as of 'one blood'.

Dynamics of tribal size

The demographic size of the Australian dialectal tribe seems to be a result of homeostatic processes. The model referred to above indicates greater social and linguistic homogeneity as the number of local groups in a tribe increases. Since the latter are fairly constant in contained numbers, this means that if tribes grow larger in population they should benefit from this internalization. Empirical evidence indicates that the dialectal tribe in Australia averages somewhere between 450 and 500 persons although the overall range is rather wide, from perhaps 200 to 2000 individuals. Tribes approaching 1000 persons show evidence of linguistic fractionating. The process may produce two or three incipient tribes, depending somewhat on the geographical configuration of the original group. So with foot mobility as a limiting factor, there is an upper numerical limit on the size of the dialectal tribe in Australia.

Should hard times reduce the size of a dialectal tribe to a point approaching 200 individuals, it shows a strong tendency to be absorbed by friendly neighbouring tribal groups. In model terms it is clear that with a reduction in the number of local groups functioning within a tribe, there is a considerable increase in the density of communications with members of other tribal groups, and of course an increase in the number of intertribal marriages. So a tribe must not fall below a certain numerical threshold if it is to maintain itself linguistically, culturally, and reproductively as a separate entity. Thus there is an optimal middle ground for the tribal population. If it becomes too large it can no longer maintain its linguistic integrity. If it is too reduced numerically, absorption is its fate.

There is evidence that with the diffusion of the initiation rites of circumcision and subincision the tribal groups which had newly adopted them fragmented in size both demographically and territorially. An analysis based upon the eastward front of these rites (Birdsell, 1973) demonstrated that with the passage of time the fragmented tribal units reintegrated and reconstituted themselves to return to normal stable numbers which characterized the Australian tribe. This recovery from the perturbation occurred as a single heavily dampened restoring action (figure 6). There was no overshoot. This perturbation and

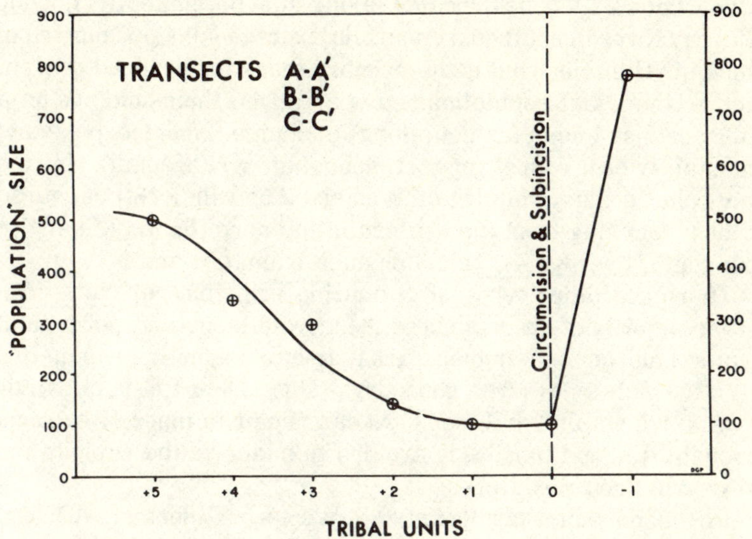

Figure 6 Restoration of tribal population numbers after perturbation due to the advance of initiation rites.

recovery are strong evidence for the homeostatic control of tribal numbers and the living space upon which they depend.

Spacing arrangements in tribal Australia

Aboriginal occupancy in precontact Australia was very tightly structured by a complex of environmental variables. The geometry of tribal holdings reflects a number of optimizing processes. In the arid stretches of the interior, where water holes and soaks are indispensable resources, tribal geometry reflects the compactness derived from minimizing the great distances to be traversed in desert country. One hundred interior tribes (Tindale, 1974) of the basic series in this analysis average 5.6 sides for their tribal boundaries. They approach a six-sided configuration, although it is irregular and far from geometrically ideal. The arrangement does seem to provide a minimum exterior boundary for a given contained space. On the other hand, 18 tribes located along the permanently flowing Murray River in southeast Australia average 4.0 sides per tribal domain. The riverfront is the primary focus for the food search, and so the tribal populations have arranged their holdings on a more or less linear fashion along its course. Not unexpectedly, coastal tribes reflect a corresponding arrangement. In the low-relief coastal country of Western Australia, 18 configured tribes, reaching from the Kariera in the north to the Minang in the south, average 4.3 sides for each tribal domain.

There are other types of optimizing principles at work. The great majority of desert tribes incorporate their most permanent water hole in their nuclear area. It is a water hole in native terms which 'has never gone dry'. This is the fall-back water hole to which all tribal members can repair in times of extreme drought. In good times it is avoided to minimize the exploitation of nearby food resources.

In the northern portion of the Western Desert of Western Australia there is a different type of configuration (Tindale, 1974). There five tribes share a single common, fall-back water hole—Karbardi. This apical water hole is equally accessible to the members of all five tribes in times of crisis. This unusual geometry reflects a dearth of permanent water holes in the region, and the adaptability of the five tribes in reaching this solution to their water needs.

Rates of intertribal marriage

The pattern of exchange of women between the local groups within tribes and between tribes reflects some degree of behavioural adaptation. Tindale (1953) surveyed the precontact matings over much of Australia, especially in the eastern, southern, and western coastal districts. He found that intertribal marriages accounted for 14.7 per cent of the total number of matings. In an unpublished survey, the writer found that for 172 precontact matings in the interior of Western Australia for which tribal identities could be accurately ascertained, the rate of intertribal marriage fell below 10 per cent (9.9 per cent). This reduction in frequency is unexpected insofar as populations living in desert areas would be expected to use the exchange of daughters to cement their rights to enter the territory of others during times of stress and water shortage. But this does not seem to be generally true. It must be concluded that the much greater distances involved between local groups and tribes in the desert acts as a greater isolating factor than it does in the better endowed coastal regions.

If the frequency of intertribal marriages within the five tribes focusing on Karbardi is examined, the data represent a small but interesting sample. Genealogical information on precontact matings is only available for four of the involved tribes and yields only nine matings. But it is suggestive that seven of these (77.8 per cent) are intertribal in character. Each of the intertribal matings involves members of adjoining tribes. It is tempting to conclude that in this unusual situation Aborigines considered it highly advantageous to exchange their women with adjacent tribes and so to minimize risks of fighting when different tribal groups were forced to refuge at Karbardi.

Asset ownership

Rights of land use for hunting and food gathering also show certain optimizing tendencies. Even in low rainfall areas local groups are still the territorial unit. They show a certain fluidity of movement and are less territorially rigid in their annual seasonal rounds than is usual in higher rainfall country. In somewhat better watered country local group holdings are well defined and occupied during more of the annual cycle. But bands visited each other during times of food surplus, to improve the variety of their diet, and of course for trading opportunities. The local group territory seemed to be a base for operations during much of the year, but friends and related groups were frequently invited into it. In the very richest country ownership of food assets was even reduced to family or individual holdings. The bunya nut trees, cycad palms, islands on which black swans flocked to nest, and a whole series of recurrent food resources were held in what might well be described as individual ownership. Again these assets were used as the basis for invitations to share in the food resources, and so to set up reciprocal networks of relationships.

Social behaviour that seems to contribute to the Australian homeostatic living system

Group behaviour in man is so varied that it is difficult to define adaptive behaviour in rigorous terms. What may seem socially beneficial in the short term may in the long run prove maladaptive. In anthropology it is currently assumed in a dogmatic fashion that most cultural behaviour is adaptive, and certainly behaviour in which changes are evident must be adaptive. This attitude short-circuits the formulation of testable hypotheses about advantageous group behaviour.

One of the problems with the identification of adaptive behaviour, as opposed to neutral or even maladaptive behaviour, is that in modern societies, in which both the demographic numbers and technology are rapidly increasing, there is no stable base against which to test either established or innovative phases of behaviour. Even among economically simple groups such as the Australian Aborigines it is by no means certain that all traditional behaviour, or even changes in behaviour, are to be considered per se as adaptive. But in a steady state society estimates of the import of group behaviour may perhaps be risked. Where social behaviour *seems* to contribute to the maintenance of a homeostatic condition it must be assumed for present purposes that it does so. The case rests upon ill-defined plausibility and is admittedly not a scientific test. In this spirit four areas of Australian social behaviour will be looked at briefly. All examples are purely qualitative and at this late date, after the total disintegration of tribal society, there seems no way of quantifying them at any future time. So in this spirit totemism will be briefly examined, as will be mechanisms for the control of the young men, the network of kinship relations, and judicial system existing among these people.

Totemism

Totemism among the Australians involves a characteristic but complex set of beliefs which enriches their lives and regulates their actions in certain directions. Totemism qualitatively certainly tends toward the maintenance of a homeostatic system. In essence totemism is a view of nature and life, of man's place in his universe, and so influences the mythologies and social groupings of the natives. It inspires their rituals and provides binding links between them, their country, and their past.

Totemism exist in many forms with varying functions. Totems may be individual in nature, dichotomized by sex, or affiliated with various types of social classes or clans. Many of its manifestations are local in character and are attached to a particular spot in the native landscape. It is connected with ceremonies which come out of the 'dream time' of the past. Totemic beliefs so strongly tie individuals to the country of their birth, indeed the spot where they are believed to be totemically conceived, that even in late postcontact times acculturated natives often strive to return to their own country as a proper place to die. The mythological bond between man and his birthplace begins at conception and is continuously reinforced throughout his lifetime by various ceremonial activities in which he participates. Upon death his spirit will return to the totemic place of his origin. The system of beliefs acts as an important stabilizer of individuals, and so of populations, in geographical space. Its influence is obviously conservative. As a spacing mechanism it acts strongly to maintain the demographic status quo.

Some aspects of totemism are connected with food taboos. If a totem refers to a given plant or animal then its members are customarily forbidden to eat it. To what extent these beliefs really do promote game or plant management has never been ascertained. They possibly serve to relieve some of the pressures of exploitation on some of the more important food products. If so, some totems will exert a stabilizing force of this type on social groups and their ecology.

Cult totems which exist over the greater part of Australia, and perhaps formerly over all of it, involve secret religious organization. They, as well as other forms of totemism, are well described by Elkin (1938, pp. 136-137).

> 'In each tribe there are a number of cult societies or cult groups or lodges, each of which consists of a number of fully initiated male members who at the same time are designated by right of birth. Each group exists to take care of and hand down a prescribed portion of the sacred totemic mythology and ritual of the tribe and often, also, to be the custodian of sacred totemic sites and to perform ceremonies for the

increase of the species which is its totem. The portion of mythology and ritual and the sacred sites entrusted to such a cult group is defined by mythological history. It is fundamentally the mythology which records the travels and actions of the tribal heroes in each subdivision of the tribal territory. The country of each local group is crossed with the paths or tracks of these heroes along which there are a number of special sites where the hero performed some action which is recorded in myth; it may have been only an ordinary everyday act, or it may have been the institution or performance of a rite.'

The significance of such cult totems for our purposes lies in the fact that they are the most common vehicle of intertribal gatherings. For example, the members of the wildcat totem may meet at a designated water hole, at a time that is advantageous seasonally, and there perform in series the secret ceremonies which describe the movements of their cult heroes along their tracks. The performers will have been born into many different local groups and perhaps into several tribes. In this fashion cult totemism forged its friendly and adaptive relations between portions of diverse local groups and tribes, and so laid the groundwork for friendly interaction at other levels. The members of a totemic lodge feel a great sense of brotherhood. They may exchange daughters as wives. This may result in the opening of an escape route should drought drive some of the cult members from their own country. Certainly at the level of simple plausibility, totemism in this and its other forms seems to be adaptive for the society and homeostatic in its actions.

Mechanisms for the control of young men

One of the means which Australian society devised to give the young girls in marriage to the old men, despite the biased sex ratio, was to defer the age of marriage for the young males. Children are generally treated with great indulgence by these people, especially young boys. But at about puberty the boys were subjected to initiation rites of various kinds before being admitted into manhood, and therefore eligible to take a wife. Throughout most of central Australia two ceremonies were involved, first circumcision and subsequently subincision. In the marginal eastern and western reaches other types of initiation ceremony were practised.

Initiation ceremonies symbolized the leaving of unfettered boyhood and the onset of strictly supervised control by the senior men of the community. Much ceremonial education occurred during the rites. A kind of subservience was imposed during the initiations. The victims were made acutely aware of the ceremonial and magical powers of their seniors. The initiations, of whatever form, were somewhat brutalizing and involved some loss of life among the initiates. Presumably this was controlled by the old men. The young men were in a sense tamed and made socially manageable.

The control of the young men continued after initiation, for they were then sent out for several years to camp on their own and subsist as bachelors in the bush. They were excluded from the camp of their own local group. This was an interval of economic testing, in which the initiates were required to be completely self-sustaining in terms of the food quest. And candidacy for marriage was so further delayed.

Finally, the young initiates remained under stringent food taboos until well into middle age. The choice parts of their kills were given in prescribed fashion to various senior relatives. Their own share was the least desirable part of the animal. Many of the choice foods were totally tabooed. In effect an intentional kind of food redistribution was practised to the advantage of their elders. Restrictions on the recent initiates were punitive, presumably educational, and socially stabilizing. Not least in consequence was the deferral of their reproductive activities for some years.

The network of kinship relations

Aboriginal society was divided into a number of classes: 0, 2, 4, and 8. These divisions structured society by placing individuals in a network of classifying kinship categories. Depending upon the number of divisions within the tribe, individuals were related to others outside of their own division in a formalistic sense, that is, not as individuals by blood descent, but in relationships specified by the network of the kinship system. These divisions also influenced marital practices.

The pattern of distribution makes it quite clear that no divisions, or simpler ones, preceded the most evolved forms. Some of the marginal tribes had no class divisions whatsoever, and among them some tribes regulated their affairs through the device of alternating generations. These usually consisted of 'those who sit in the sun' and 'those who sit in the shade'. This simple dichotomy served to regulate their social affairs well. In a somewhat less marginal position were tribes with a two-class or moiety system. They divided society into two halves, but not specifically along alternating generational lines. Moieties are generally considered devices for integrating societies. Marriage was outside of one's moiety. The network of relationship was determined by moiety affiliation.

Further toward the interior tribes were commonly divided into four classes or sections. An individual could only marry persons in one of the three sections other than his own. So a man was limited to obtaining a wife from one-quarter of the available women. This rather restricted a man's choice of wife and so required a search over a greater area to find a suitable bride for whom he was eligible. This was the predominant system in precontact Australia.

Originating in the north-central portion of the continent was the eight class or subsection system. It seems to have been spreading and overriding adjacent, less elaborate systems at the time when White colonization disrupted tribal society. It restricted a man's search for a woman to only one of the seven subsections other than his own. Aran Yengoyan (1968) has hypothesized that the more complex class systems were adaptive from the very fact that they required a more extended search to find a wife. Thus friendly contacts were developed with more distant local groups. This far-flung network of relationship would also increase the likelihood of finding sanctuary further away in times when one's own waters have gone dry. The idea is certainly plausible.

Perhaps standing somewhat against Yengoyan's hypothesis is the fact that some desert tribes such as the Pitjandjara and Ngalea functioned with no classes whatsoever in societies characterized by alternating generations. They managed to maintain themselves successfully in very stringent domains. Again, the rate of intertribal marriage in the Western Desert of

Western Australia, as indicated earlier, falls well below the continental average. These data may indicate that the network of relationships was not as extensive there as in the more densely settled and better-watered regions. So it cannot be determined now how much the elaboration of class systems in Australia was a matter of sheer innovation for the sake of something new, or whether these divisions really were designed to introduce stabilizing factors in the small societies of the more arid reaches.

The judicial system

In spite of their simple economy the Australians evolved a flexible and very suitable judicial system. They had a well codified system of punishment. It took the form of trial by ordeal for most offences against individuals and groups. This is not the place to detail the ordeals but they varied judiciously with the seriousness of the crime. Since there was no strong individual leadership, the level of ordeal was determined after prolonged discussion among the seniors of both sides of the dispute. The literature implies that only men were involved in this decision making, but the role of women has notoriously been underreported for these people, and it may well be they added to the group consensus.

The offender was made to stand before a group of those representing the injured party and his kin, and to submit to a shower of spears, boomerangs, and throwing sticks. The distance from which the weapons were thrown, their timing, number, and type was determined by consultation. In most cases the offender was at least morally backed by members of his own group, and at times a person, such as a brother or wife, was allowed to stand alongside to help parry the missiles. These affairs were usually terminated with the first blood-drawing. Since men had practised dodging weapons from earliest childhood, injuries were not apt to be severe unless a great number of weapons were thrown simultaneously and from close range. In short, the ends of justice could be loaded. Thereafter the groups would resume friendly relationships with all ill-feelings formally dissipated. These forms of social outlet for anger served a highly stabilizing function in the intergroup relationships between members of different bands. It was not a bad system.

There is even some suggestive evidence that devices for adjudicating land disputes between local groups existed among these people. Since a demographic steady state could not be maintained in every local population, it is obvious that some form of adjustment would be required to stabilize territorial rights. In one instance in New South Wales, Frazier (1892) reports on such a dispute. One local group prospered mightily, increasing its numbers beyond the optimum carrying capacity of its territory. An adjacent band had declined numerically. The members of the more numerous band proposed to take the matter in their own hands by force and simply expropriate some of the territory of the smaller group. Since a number of the tribal bands were assembled in one place, the argument was transferred to them as a whole. Finally a decision was reached on how best to settle the dispute—by combat between single champions of the bands. Ironically the result was not reported by Frazier, but it is of interest that social mechanisms existed to handle such cases.

Another instance that occurred on the Coorong of coastal south-eastern South Australia comes from N. B. Tindale (personal communication). There, a landowning local group left its territory and moved into the new White settlement of Kingston. This left their band territory unoccupied. Further up the Coorong a local group which had been continuously harassed by their neighbours petitioned the assembled members of the tribe for permission to occupy the now vacated land. The request was granted so the quarrelling local neighbouring groups were separated. Both of these examples occurred in early postcontact times, but they do show a framework within which land disputes could be settled and so make accommodation for the waxing and waning populations of bands. The homeostatic import of such devices is self-evident.

Conclusions

An ecological analysis has demonstrated with its bivariate high coefficient of correlation that aboriginal densities were tightly controlled throughout Australia by the variables of the environment. The man-land relationships were tightly structured. Population biology was manipulated so as to reinforce the limitations imposed by the environment. Preferential female infanticide was extensively practised and involved the death at birth of at least 15 per cent of the newborn. There is evidence that this systematic murdering occurred within a framework of conscious population regulation. The three basic units in aboriginal society were the family, the local group, and the tribe. All three of these units evidence homeostatic forces tending to produce optimal numbers. Finally four aspects of social behaviour: totemism, initiation rites for young men, the network of kinship relations, and the judicial system all plausibly operated to maintain a stable living system among these economically simple people.

References

Birch, L. C. (1948) The intrinsic rate of natural increase of an insect population. *Journal of Animal Ecology*, **17**, 15–26.

Birdsell, J. B. (1953) Some environmental and cultural factors influencing the structuring of Australian aboriginal populations. *The American Naturalist*, **87** (834), Supplement, 171–207.

Birdsell, J. B. (1957) Some population problems involving Pleistocene man. *Cold Spring Harbor Symposium on Quantitative Biology*, **22**, 47–69.

Birdsell, J. B. (1958) On population structure in generalized hunting and collecting populations. *Evolution*, **12** (2), 189–205.

Birdsell, J. B. (1968) Some predictions for the Pleistocene based on equilibrium systems among recent hunter-gatherers. In *Man the Hunter*, ed. Lee, R. B. and DeVore, I. pp. 229–240. Chicago: Aldine Press.

Birdsell, J. B. (1973) A basic demographic unit. *Current Anthropology*, **14** (4), 337–356.

Cudmore, A. F. (1893) A wild tribe of natives near Popiltah, Wentworth, New South Wales. *Australian Association for the Advancement of Science*, **5**, 524–526.

Curr, E. M. (1886) *The Australian Race*, Vol. II. p. 46. Melbourne: Government Printer.

Elkin, A. P. (1938) *The Australian Aborigines*. Sydney: Angus and Robertson.

Frazier, J. (1892) *The aborigines of New South Wales*. Sydney.

Kryzwicki, L. (1934) *Primitive Society and its Vital Statistics*. London: Macmillan.

Martin, C. R. and Reed, D. W. (1976) The relation of mean annual rainfall to population density for some African hunters and gatherers. In *The Perception of Evolution*, ed. Mai, L. Anthropology UCLA 7 (in press).

Tindale, N. B. (1940) Distribution of Australian aboriginal tribes: a field survey. *Transactions of the Royal Society of South Australia*. **64**, 140–231.

Tindale, N. B. (1953) Tribal and intertribal marriage among the Australian aborigines. *Human Biology*, **25** (3), 169–190.

Tindale, N. B. (1974) *Aboriginal Tribes of Australia*. Berkeley: University of California.

Vorkapich, M. (1976) Population density and Great Basin ecology. In *The Perception of Evolution*, ed. Mai, L. Anthropology UCLA 7 (in press).

Yengoyan, A. (1968) Demographic and ecological influences on aboriginal Australian marriage sections. In *Man the Hunter*, ed. Lee, R. B. and DeVore, I. pp. 185–199. Chicago: Aldine Press.

Adaptive strategies, late Palaeolithic life styles: Towards a theory of preventive medicine

VALERIUS GEIST
Faculty of Environmental Design, University of Calgary, Calgary, Alberta, Canada

Introduction

The theme of this symposium is social behaviour as a means of controlling resources and thereby directly or indirectly controlling the number of individuals in populations. If we are interested in gaining broader insights into population regulation, however, with a view to contributing to the solution of the human population crisis, we can profitably ask the converse, namely, how resources—more precisely, strategies of resource exploitation or economic strategies—control social behaviour and reproduction. The justification for this is not only that the latter approach has historically been most productive in explaining adaptive syndromes in animals—not only social systems or social behaviour—but also because it promises to be equally productive when applied to human beings. Consider the implications of the proposition that a society's moral values, language, social institutions, forms of conduct, policies, and behaviour towards its neighbours are largely a result of its economic strategy. The inescapable corollaries are that social values are bent or broken as new economic strategies succeed old ones, that diversity of economic strategies may lead to the precarious coexistence of contradictory social values, and that to maintain cherished values it is essential to control economic strategies to make them compatible with social values.

The relation between ecological determinants and social adaptations in animals was demonstrated first in a flurry of publications based on the study of birds, such as those of Cullen (1957), Brown (1964), Crook (1965), Verner and Willson (1966), and Lack (1968). Almost simultaneously, and quite independently, appeared Eisenberg's (1966) pioneering work on the social organization of mammals. Thereafter a great many studies reported in our classical journals or symposia showed correlations between social adaptations and ecological variables, to be summarized recently in Wilson's (1975) synthesis on sociobiology.

In my own work, I broke away from the search for correlations between ecological variables and behaviour and concentrated on exploring the logical relationship between ecological strategies and the consequent adaptive syndromes. The latter I define as the constellation of diverse adaptations that are part and parcel of an adaptive strategy. Using this approach, I explained for ungulates various aspects in the evolution of social systems, communication, and sexual dimorphism (Geist, 1967, 1974a, 1974b) and, more recently, the ecological determinants shaping the morphology of weapons and defences, as well as strategies and tactics of aggression (Geist, 1977, in press). The same approach permitted me also to show that all characteristics of the class Mammalia, that is, those features which we share with all mammals, can be deduced without contradiction from one ecological niche, namely, that of a nocturnal small-bodied reptillian carnivore from temperate climates (Geist, 1972). This illustrated the utility of the concept of adaptive syndrome as a product of ecological determinants in reconstructing life forms from the past.

About six years ago, I addressed myself to the question of how to maximize health in human beings through environmental design. This led to an examination of humans from different environments, past and present, because it became evident earlier that phenotypes reflect the environment they develop in and that certain phenotypes were healthier than others. It was in the subsequent work that I became aware that some anthropologists, sociologists, and philosophers were interpreting their findings in much the same way as I was, be it the chilling account by Turnbull (1972) of the Ik, the analysis by Stover (1974) of the forces shaping Chinese civilization, Carneiro's (1970) views on the ecological determinants of nation states, or Cliff Hooker's (1976) preliminary account of his analysis of the modern market society. Here appears to lie the hiatus between the biological and social sciences and probably the way to the grand synthesis Wilson (1975) anticipates.

Towards a theory of health

The quest for a theory of health confirmed the utility of conceiving of ecological variables as determinants in the social organization of human beings, but it resulted in a rather different theory of human evolution from that which is broadly accepted today. A theory of human evolution, however, appears fundamental to a theory of preventive medicine. The reasons for this conclusion are as follows.

The concept of reproductive fitness predicts that in order to maximize fitness, individuals ought to expand into unoccupied habitats or into unused resources whenever possible. Individuals are thus likely to be programmed to exploit opportunities to disperse. However, upon entering regions previously unoccupied by the species, colonizing individuals encounter conditions that are unusual: they are faced with an abundance of food unexploited by conspecific competitors; they encounter new types of food, climates, landscapes, predators, competitors, parasites, and diseases; their few conspecific competitors—for reasons outlined below—will differ considerably in their social behaviour from those in their parent population. Dispersing individuals must, therefore, differ from individuals in the parent population by being able to cope with the new and unforeseeable. They must be able to shape their behaviour so as to acquire and use a maximum of information. They must be roamers and explorers. They must have a capable immune system. They must be at the outer limit of muscular power and endurance possible to their specific genome, and if this is so then they must deal in combat with far more powerful conspecific foes. They must be able to suppress appetites for unadaptive old behaviours and suppress aversions against adaptive new ones, and we expect them to reproduce maximally. In short, we expect a 'dispersal phenotype' during dispersal and a 'maintenance phenotype' in stable old populations at carrying capacity. The notion of an 'r-phenotype' and 'K-phenotype', following a popular conception in population biology, would be too simplistic.

In our studies with mountain sheep we noticed that different environments produced different phenotypes. Those individuals raised under very favourable resource conditions differed significantly in behaviour and morphology from those raised under adverse conditions; we termed them 'high-quality' and 'low-quality' individuals respectively (Geist, 1971; Shackleton, 1973; Horejsi, 1975). Populations of high quality were characterized by individuals of exceptional body size and growth rates, early maturation, very good maternal care, high reproduction rate, a high frequency of play and exploration, frequent and intense social interaction, and generally a striking vigour of behaviour. Peculiarly, they had also a short life expectancy. Morphologically they were characterized by exceptional development of tissues of low growth priority. Subsequent reviews of the literature added to the syndrome: high-quality individuals appear to have few congenital deformities, a low rate of parasitism, strong skeletons, and a high haematocrit. It was only later that we recognized that we were dealing here with a functional phenotype and renamed it the 'dispersal phenotype'. We are dealing with the phenomenon of genes shaping the individual in accordance with signals received from the environment.

When dispersing into new environments, individuals must be able to shape new behaviours to deal with new problems. Clearly, some are likely to be better endowed genetically to do this than are others. If all individuals are forced to solve the same problems and exert themselves maximally to do so, those better endowed genetically to solve the problems have the higher fitness. Note, individuals have epigenetic mechanisms at their disposal to adapt phenotypically. Only when these are used maximally (with environmental variance, therefore, approaching zero) are the underlying genetic differences between individuals exposed to natural selection (genetic variance being maximized, approaching unity). Therefore, during dispersal into previously unoccupied habitat there is a rapid genetic rearrangement favouring those traits that are linked to successful coping with the new problems. The individuals live at that time in an evolutionary environment, which can be defined as that which maximizes phenotypically (and later genotypically) specific adaptive traits. These traits will be termed 'diagnostic features'. Thus, the evolutionary environment of a species is that which maximizes phenotypically the diagnostic features of a species.

To maximize adaptations of a morphological, behavioural, or

physiological nature requires energy and nutrients; it is best done under conditions of resource abundance. We noticed earlier that phenotypes produced under this condition were those of the greatest health. They were not only healthy physiologically, but also could—in theory—cope best with new problems and therefore remain healthy. Therefore, the environment of maximum health must be equal to the evolutionary environment. Here is the link between evolutionary biology and preventive medicine.

We can explore the environment that maximizes health by defining what the diagnostic features of our species are, features which differentiate us from our parent species, and then search for the factors which will maximize these. In view of the difficulty of defining diagnostic features, this is not a simple matter, but it is not a hopeless one. A second approach can also be used, to look for exceptional phenotypic development in human beings and reconstruct their environments. This approach quickly brings to the attention the late Palaeolithic people which reached a development unequalled by any population since. They were not only tall, athletic, robust, and remarkably free of evidence of disease (Vallois, 1961; Weidenreich, 1939; Howells, 1973), but they also had brains larger than modern populations, the largest average brain sizes in fact (Mettler, 1955; Montague, 1960; Hulse, 1963), and they were virtually free of cannibalism and homicide. This may be contrasted with the findings from the Mesolithic, the first dark age of modern man (see Clark, 1965; Heinzelin, 1962; Angel, 1968; Armelagos, 1969).

It is not only people from the late Palaeolithic who satisfy the morphological criteria of the dispersal phenotype, namely, that which has well-developed tissues of low growth priority; there are others. Herders and fishermen compared to plant agriculturists tend to show it, and so do classes within populations. Individuals from upper classes tend to differ from those of lower classes by virtue of their better developed bodies (Tanner, 1962; Huber, 1968). Therefore, one strategy of research, not undertaken yet, would be to compare the life styles of those approaching a dispersal phenotype against those that do not, and from the similarities and differences develop a testable theory of how to maximize health environmentally. As a first step, however, one would have to decipher the adaptive syndrome, life styles, or economic strategies of the late Palaeolithic era.

To understand an adaptive syndrome, it is necessary first to understand the one it is derived from. To understand the mammalian system, one must begin with an understanding of the reptilian one and trace the steps in the transformation and the conditions causing them. In a similar vein, to understand modern man's adaptive syndrome—what accounts for its diagnostic features—one must know the adaptive syndrome of the parent species, and to understand that we must unravel it in turn. I did this and, not surprisingly, the history of our genus is quite similar to that of other ice-age mammals in that it begins with tropical forms and radiates to progressively more unstable, extreme, and demanding environments until it stops to take its final form in the periglacial environments of the northern latitudes.

Second, one must understand periglacial environments. One can reconstruct these provided the ecology of northern terrestrial ecosystems and the autecology of large northern mammals are understood, and the remnants of periglacial ecosystems available today have been examined.

Third, to decipher late Palaeolithic life styles, one must understand the principal food of late Palaeolithic people, that is, large grazing mammals, and their biology.

Fourth, one must have a theory of phenotype development in large mammals, much of it developed in animal science, to grasp one very striking peculiarity of the late Palaeolithic populations: they maintained an exceptional development of individuals throughout their long span of existence. This can be done only if the resources available for growth and reproduction are expanding as fast as or faster than the population—and no such environment is conceivable, let alone known. Conversely, the late Palaeolithic cultures could curtail fertility artificially. The reason for this became apparent when I analysed the adaptive strategies of periglacial hunters, for it became evident that to exist in the periglacial zones the human cultures there had to maximize the physical, intellectual, and social competence of individuals; they had to maintain individuals artificially in the 'dispersal phenotype' at the expense of population size.

The foregoing is a sketch of one part of a book I have prepared for publication. Some of it has surfaced, albeit in rather sketchy form in a smaller book (Geist, 1975a). I presented a few of my findings in one paper (Geist, 1975b) and in a brief article on periglacial ecology in *Nature Canada* (Geist, 1975c).

Late Palaeolithic economic strategies

The adaptive strategies of the late Palaeolithic hunters from the periglacial zones of Europe and the Mediterranean basin appeared to revolve around two principal methods of obtaining food from large mammals. These were the only food source during the eight or more months of winter. Because of the need to wear clothing, and the fact that silent stalking is impossible in snow and in clothing, the hunters could only hunt cooperatively. Thus, cold and snow as ecological determinants limit the alternatives in hunting methods.

The first strategy of killing large animals I call long-distance confrontation hunting. It can be visualized as similar to bull-fighting, but carried out by a closely cooperating team of men varying their methods with the species and conditions of the hunt. Prerequisites for this type of hunting are superb physical fitness, strength, and coordination, as the hunter must be able to approach, and then evade the attacks of the quarry while focusing the attention of the quarry on himself in order to relieve an endangered companion. It requires close cooperation between individuals, good judgement so as to choose only such a quarry in such a situation as will minimize danger to the hunters, and familiarity with the behaviour of all the large mammals hunted—that is, of each sex-age class, under the different situations encountered, at any time in the yearly cycle. It required unanimity in decision-making and altruism to make this strategy operate. As a weapon, a narrow-bladed spear is required to penetrate deeply when thrown, to maximize haemorrhaging by maximizing the length of the wound channel. This leads to the shortest interval between the quarry being struck and its death, thus minimizing danger. Long-distance confrontation hunting can take place only on hard unobstructed ground, for otherwise the manoeuvrability of the hunters is impeded. Its success is based on the readiness of large mammals to stand and defend their young against predators or to attack predators. Neanderthal man, who will not be discussed here, apparently practised short-distance confrontation hunting, and the adaptations of this highly specialized human form are derived without contradiction from it. In short-distance confrontation hunting, the hunter reduces the danger to himself by attaching himself to his quarry as a means of evading its attacks.

The long-distance confrontation hunting method was probably an important, but not the most important, method of obtaining food. The most important method was to intercept migratory herds of ungulates, kill more than immediately needed, convert this excess into a storable product, and use it to supplement the deficits of hunting methods that were exercised when interceptions of herds could not be practised, in winter. Ungulates tend to migrate in fall and in spring, and that is when they can be intercepted. Prerequisites for this strategy of exploiting ungulates for food are as follows. An ability to keep track of chronologic time accurately throughout the year is necessary, for ungulate movements cannot be accurately predicted by weather phenomena or seasonal changes in the landscape: a lunar calendar, as evidenced by Marshack's (1972, 1976) analysis of the bâton de commandement, fulfils this requirement admirably. Interception hunting required the skill to turn masses of meat quickly into a storable form. This, in turn, required a method of rapidly cutting the meat across the grain into thin slices, and drying it quickly despite a threat of rain; it also required a very large amount of labour lest the precious meat spoiled. The blade industry of the late Palaeolithic satisfied, in large part, the first requirement, as it was capable of producing quickly, and economically in raw materials, a large number of very sharp backed blades with straight edges for accurate cutting. The increase in the number of fires in the Palaeolithic, fires characterized by wood ash rather than bone ash (Oakley, 1961; Pfeiffer, 1972, p. 233), indicates that such fires were probably often used for drying meat, as well as for imparting smoke to it as a preservative. The need for the labour of all members of the group predicts the robust bones and muscle insertions typical of the skeletons of late Palaeolithic women, and it predicts that older individuals could enhance the reproductive fitness of their children through their labour. This opens the door to the existence of the closely knit extended family.

Once a large amount of meat was dried, it had to be stored in localities where it could not spoil and would be protected from large carnivores. This predicts the development of shelters and microclimates favouring the preservation of dried meat, no matter what the weather was, and it predicts the banding together of several families to protect their resources. There is evidence for both (Clark, 1970; Klein, 1973).

In this adaptive syndrome we can identify three major worries of the male. The first was to consistently encounter and then capture game without being injured. In particular, the correlation between the paths of the moon across the night sky and the appearance of migratory game herds must have been puzzling. Let it be clear: in this adaptive syndrome the hunter does *not* follow game herds; he intercepts them, kills, and watches them disappear, not encountering them again for many months. If he fails to be at the right place at the right time, his family may starve. His concerns about his food are reflected in his preoccupation with its images, as recorded in late Palaeolithic art.

His second very big concern was with women. In the late Palaeolithic adaptive syndrome, the male, probably for the very first time, depended on woman for a successful life. She supplied essential services such as the maintenance of shelter and clothing in a strikingly harsh inhospitable landscape. He had to provide materials and also create conditions in which the female could perform such services to himself, the children, and the few old individuals. Yet the priorities of male and female probably diverged so that discord could develop over when to move camp and thus disrupt the security of it. Such moves had probably to be undertaken when stored food was still ample and this only added to the labour of everyone concerned. Yet we know that serious discord would be intolerable, for it would generate psychological stress and such stress would likely impair both lactation and foetal development (Nuckolls *et al.*, 1972; Newton and Newton, 1950), reducing the competence and development of children. Yet that had to be avoided at all cost, because only by maximizing the physical, intellectual, and social competence could the individual hope to live and succeed in the periglacial zones. The interest of parents in highly competent children was likely to be substantial, as their life expectancies, for the first time, were such that they could expect to grow old and in need of support (Vallois, 1961; Cook, 1972). The male could, therefore, not impose his will directly, because a supportive social milieu conducive to lactation and the children's ontogenetic development had to be maintained at all times, but only indirectly, probably by a cult. This is what the female figurines of the Venus cult suggest (as well as an analysis of their postures). Clearly, when a male died he needed both his weapons in his grave and his bâtons de commandement, to ensure success in his afterlife, and his Venus figurines to ensure an obedient wife and a harmonious family.

The concept that, in the late Palaeolithic, people maximized the individual competence of individuals by any and all means is based on the following evidence. The periglacial environment was the most extreme which had been colonized by human beings until then, and this demanded the greatest competence of individuals to deal with diverse matters. It was necessary to succeed physically in long-distance confrontation hunting, which demanded great agility, coordination, strength, and discipline. Strength and endurance would also be required for moving heavy equipment and food stocks when the camps were moved. Endurance would also be a prerequisite for lengthy roaming, now permitted by the availability of dry meat obviating the need for killing; endurance, however, is, in part, a function of body size. There was need for great intellectual competence to outdo strategically and tactically some 15–20 species of large mammals at all seasons of the year, and for a great deal of knowledge to exploit the resources of a winter landscape, not only those offered by large mammals. It was necessary to be competent in social interactions in order to maintain cooperation with other hunters, to teach children, to maintain a harmonious family milieu, to arrange the time of departure of camps, prepare for cooperative hunts, distribute work and the results of the hunt equitably, act in a manner to ensure group harmony by minimizing discord lest the abilities of any of its members be undermined, to find a successful link to another group and integrate the whole family with it if disaster befell the one group, particularly in hard times. By himself, an individual could not survive—not with late Palaeolithic technology—but had to depend on the acceptance, goodwill, and cooperation of others.

If maximization of individual competence was the goal of late Palaeolithic hunters, they not only had to feed well on proteinaceous food, provide a harmonious social milieu to maximize ontogenetic development, and provide diverse exercises that permitted individuals during ontogeny to gain a knowledge of, and then competence in, the many diverse physical skills, knowledge, and rituals, but they also had to maximize heterosis. This they could not do without a system of kinship. They had to keep track of who mated with who, lest precious resources were wasted and dangers courted by producing less than fully competent individuals.

They also had to somehow control the number of progeny, reducing the number born, and spacing the remainder so as to maximize their ontogenetic development. How they did this we do not know, of course, but if they did maximize individual development then they had to space births.

The adaptive strategies sketched out above explain, in addition, the adaptive functions of such acts as dancing (a means of maintaining precious physical fitness between episodes of confrontation hunting) and music (a means of facilitating dancing by numbing physical discomfort and of facilitating social cooperation) and poetry (as a means of facilitating recall of large amounts of verbal information and passing it on without significant distortion from generation to generation) and they explain differences in sexual organ structure between the Kohisan and other races (see Geist, 1975a, p. 190).

If the late Palaeolithic people did participate in very diverse physical and intellectual activities, as the above adaptive strategies and their consequences indicate, then experimental evidence of the effect of environment on brain size predicts that they ought to have had exceptionally well developed and large brains (Cummins *et al.*, 1973; Zimmerberg *et al.*, 1974; Greenough, 1975). This they had indeed (Mettler, 1955; Montague, 1960; Hulse, 1963). There is also little evidence of cannibalism and homicide, and not surprisingly: individuals, be they male or female, were precious to the group and exposure to hazards beyond those encountered normally would bring hardship to everyone. Intergroup and intragroup aggression was, therefore, quite intolerable. Moreover, the adaptive strategies of the late Palaeolithic could not result in large amounts of stored surplus resources acting as an incentive to raids; the resource they primarily relied upon, large mammals, were very mobile and not a defendable resource localized in space. The conditions for warfare to flourish simply did not exist. There was neither social status nor material possession to be gained by aggression.

In maximizing individual competence, the late Palaeolithic people artificially maintained and maximized the 'dispersal phenotype'. If this is valid, then the late Palaeolithic people ought to have dispersed at an unprecedented scale over areas previously not colonized. Indeed, beginning with the late Palaeolithic, one does find evidence for waves upon waves of people moving from the Eurasian continent to settle all others. By the end of the last glaciation, they had colonized Australia, America, and much of Africa, and had begun to adapt to local conditions and new adaptive strategies.

The model of late Palaeolithic man as sketched out above can be verified by its predictions about health. Each major element of the postulated late Palaeolithic life style ought to have led to superior health, as well as to superior physical, intellectual, and social competence following ontogeny. Conversely, if we search for factors that enhance intellectual development, or physical development, or the ability of an individual to cooperate with others, then in each case we ought to wind up with a constellation of factors each of which mirrors the late Palaeolithic life style. This is essentially a theory of preventive medicine. So far, I have not found evidence to contradict this view, but on the contrary. Each major (and even minor) characteristic postulated for the late Palaeolithic life style is associated with superior health or individual competence, such as a bulky high-protein diet, the supportive social milieu provided by an extended family and groups of closely cooperating individuals, frequent diverse and intense physical exercise, acquired intellectual competence or life in natural surroundings and thus exposure to sunshine, natural sounds, unadulterated air, and the stimuli of diverse climatic and habitat factors.

A number of implications follow from the postulated late Palaeolithic model. It indicates that we as a species, by artificially supporting ourselves in a dispersal phenotype, are essentially a 'self-made' species, that population and eugenic controls are nothing novel but as old as 'humanity' itself, that natural and evolutionary environments are quite different and must not be confused, and that since all human races of today are earlier or later descendants of northern hunters the present-day hunter-gatherers are secondarily—not primarily—'primitive'. Such societies are not at all old, nor indigenous, do not reflect life in an evolutionary environment, and are not very useful as models of life styles maximizing health, contrary to Boyden's (1972) suggestion.

By inverting the theme of our symposium and applying it to humans during the last major evolutionary phase, when individuals reached exceptional phenotypic development, we do indeed note a relationship between population regulation and resources. However, it is not of the type indicated by our theme, nor is there any positive relation here between violence and resources, but the reverse. Throttling reproduction by the reduction of resources available for reproduction and growth would have led to phenotypically poorly developed, frail, sickly individuals with small brains, abnormal behaviour, and probably an aggressive disposition. We suspect this from animal studies, and also from studies of malnourished and undernourished human beings, as evidenced in studies such as Tanner's (1962), Alexander's (1961), Alexander and Peterson's (1961), Schneour's (1974), Levitsky and Barnes's (1972), Williams's (1971), or in volumes such as that edited by Scrimshaw and Gordon (1968). We suspect that in many large mammals, population control is achieved through neonate mortality under conditions of resource shortage, and is associated with considerable phenotype plasticity. I can only conclude that human beings must find their own means of controlling populations and adjusting them to available resources. Should we, by default or design, fall back on 'natural' means of population control, such as that noted above, then the terrible price will be the loss of individual development and competence, and a cult of poverty akin to that described by Turnbull (1972) for the Ik—or maybe much worse.

References

Alexander, G. (1961) Energy expenditure and mortality in new-born lambs. *Proceedings of the 4th International Congress on Animal Production*, **3**, 630–637.

Alexander, G. and Peterson, J. E. (1961) Neonatal mortality of lambs. *Australian Veterinary Journal*, **37**, 371–381.

Angel, J. L. (1968) Ecological aspects of paleodemography. In *The Skeletal Biology of Earlier Human Populations*, ed. Brothwell, D. R. pp. 263–270. Oxford: Pergamon Press.

Armalagos, G. J. (1969) Disease in ancient Nubia. *Science*, **163**, 225–259.

Boyden, S. (1972) Biological determinants of optimum health. In *Human Biology of Environmental Change*, ed. Vorster, D. J. M. London: International Biological Programme.

Brown, J. L. (1964) The evolution of diversity in avian territorial systems. *Wilson Bulletin*, **76** (2), 160–169.

Carneiro, R. L. (1970) A theory of the origin of state. *Science*, **169**, 733–738.

Clark, J. D. C. (1965) Primitive man in Egypt, West Asia and Europe in Mesolithic times. In *Primitive Man in Egypt, West Asia and Europe*, ed. Garrod, D. A. E. and Clark, J. G. D. pp. 23–61. Cambridge: Cambridge University Press.

Clark, J. D. (1970) *The Prehistory of Africa*. Southampton: Thames and Hudson.

Cook, S. F. (1972) Aging of and in populations. In *Developmental Physiology and Aging*, ed. Timiras, P. S. pp. 581–606. New York: Macmillan.

Crook, J. H. (1965) The adaptive significance of avian social organization. *Symposia of the Zoological Society of London*, **14**, 181–218.

Cullen, E. (1957) Adaptations in the kitiwake to cliff-nesting. *Ibis*, **99** (2), 275–302.

Cummins, R. A., Walsh, R. N., Budtz-Olsen, O. E., Konstantinas, T., and Horsfall, C. R. (1973) Environmentally-induced changes in the brain of elderly rats. *Nature, London*, **243**, 516–518.

Eisenberg, J. F. (1966) The social organization of mammals. *Handbuch der Zoologie*, **10** (7), 1–92.

Geist, V. (1967) A consequence of togetherness. *Natural History*, **76** (8), 24–31.

Geist, V. (1971) *Mountain Sheep: A Study in Behaviour and Evolution*. Chicago: University of Chicago Press.

Geist, V. (1972) An ecological and behavioural explanation of mammalian characteristics, and their implication to Therapsid evolution. *Zeitschrift für Saugetierkunde*, **37**, 1–15.

Geist, V. (1974a) On the relationship of social evolution and ecology in ungulates. *American Zoologist*, **14**, 205–220.

Geist, V. (1974b) On the relationship of ecology and behaviour in the evolution of ungulates: Theoretical considerations. In *The Behaviour of Ungulates and its Relation of Management*, Vol. 1, ed. Geist, V. and Walther, F. pp. 235–246. Morges: International Union for Conservation of Nature.

Geist, V. (1975a) *Mountain Sheep and Man in the Northern Wilds*. New York: Cornell University Press.

Geist, V. (1975b) About natural man and environmental design. In *Science and Absolute Values*, Vol. 1. Proceedings of the Third International Conference on the Unity of the Sciences, London, 1974.

Geist, V. (1975c) On life in the sight of glaciers. *Nature Canada*, **4** (3), 10–16.

Geist, V. (1977) On weapons, combat, and ecology. In *Aggression, Dominance and Individual Spacing*, ed. Krames, L. Advances in the Study of Communication and Effect, Vol. 4. New York: Plenum Press.

Greenough, W. T. (1975) Experimental modification of the developing brain. *American Scientist*, **63**, 37–46.

Heinzelin, J. de (1962) Jshango. *Scientific American*, **206** (6), 105–116.

Hooker, C. (1976) Cultural form, social institution, physical system: Remarks towards a systematic theory. In *Man and his Environment*, ed. Mohtadi, M. F. Vol. 2, pp. 169–182. Oxford: Pergamon Press.

Horejsi, B. L. (1975) Suckling and feeding behaviour in relation to lamb survival in bighorn sheep. Ph.D. dissertation, University of Calgary.

Howells, W. (1973) *The Pacific Islanders*. People of the World Series. London: Weidenfeld and Nicolson.

Huber, N. W. (1968) The problem of stature increase: looking from the past to the present. In *The Skeletal Biology of Earlier Human Populations*, ed. Brothwell, D. R. pp. 67–102. Oxford: Pergamon Press.

Hulse, F. S. (1963) *The Human Species*. New York: Random House.

Klein, R. G. (1973) Ice-age hunters of the Ukraine. *Prehistoric Archeology and Ecology*. Chicago: University of Chicago Press.

Lack, D. (1968) *Ecological Adaptations for Breeding Birds*. London: Methuen.

Levitsky, D. A. and Barnes, R. H. (1972) Nutritional and environmental interactions in the behavioural development of rats: long-term effects. *Science*, **176**, 68–71.

Marshack, A. (1972) Upper Paleolithic notation and symbols. *Science*, **178**, 817–827.

Marshack, A. (1976) Implications of the Paleolithic symbolic evidence for the origin of language. *American Scientist*, **64**, 136–145.

Mettler, F. A. (1955) Culture and the structural evolution of the neural system. In *Culture and the Evolution of Man* (1962), ed. Montague, M. F. A. pp. 155–201. New York: Oxford University Press.

Montague, A. (1960) *An Introduction to Physical Anthropology*. Springfield, Ill.: Charles C. Thomas.

Newton, N. and Newton, M. (1950) Relation of the let-down reflex to the ability to breast feed. *Pediatrics*, **5**, 726.

Nuckolls, K. B., Kaplan, B. H., and Cassel, J. (1972) Psychosocial assets, life crisis and the prognosis of pregnancy. *American Journal of Epidemiology*, **95**, 431–441.

Oakley, K. P. (1961) On man's use of fire, with comments on toolmaking and hunting. In *Social Life of Early Man*, ed. Washburn, S. L. pp. 176–193. Chicago: Aldine Publishing.

Pfeiffer, J. E. (1972) *The Emergence of Man*. 2nd edition. New York: Harper and Row.

Scrimshaw, N. S. and Gordon, J. E., eds. (1968) *Malnutrition, Learning and Behaviour*. Cambridge, Mass.: MIT Press.

Shackleton, D. M. (1973) Population quality and Bighorn Sheep, Ph.D. dissertation, University of Calgary.

Shneour, E. (1974) *The Malnourished Mind*. Garden City, New York: Anchor.

Stover, L. E. (1974) *The Cultural Ecology of Chinese Civilization*, New York: Mentor New American Library.

Tanner, J. M. (1962) *Growth at Adolescence*, 2nd edn. Oxford: Blackwell.

Turnbull, C. M. (1972) *The Mountain People*. New York: Simon and Schuster Touchstone Book.

Vallois, H. V. (1961) The social life of early man: the evidence from skeletons. In *Social Life of Early Man*, ed. Washburn, S. L. Chicago: Aldine Publishing.

Verner, J. and Willson, M. F. W. (1966) The influence of habitats on mating systems in North American passerine birds. *Ecology*, **47**, 143–147.

Weidenreich, F. (1939) The duration of life of fossil man in China and the pathological lesions found in his skeleton. *Chinese Medical Journal*, **55**, 34–44.

Williams, R. J. (1971) *Nutrition against Disease: Environmental Prevention*. New York: Pitman.

Wilson, E. O. (1975) *Sociobiology: The New Synthesis*. Cambridge, Mass.: Harvard University Press/Belknap.

Zimmerberg, B., Glick, S. D., and Jerussi, T. (1974) Neurochemical correlate of a spatial preference in rats. *Science*, **175**, 623–625.

Introductions and discussion

Session One

Chairman: F. J. EBLING

Chairman's introduction

To open this symposium is a privilege which I cannot claim of right but which I appreciate. While the gestation has largely been mine, the conception was that of Dr Michael Stoddart, who has just returned from a year's sojourn in the Antipodes. I welcome him back.

Since I first read *Animal Dispersion in Relation to Social Behaviour* I have felt that it must be one of the greatest works in biology since *On the Origin of Species*. It is, indeed, a very much easier book to read, for by modern standards Darwin was by no means succinct as a writer, whereas Wynne-Edwards neatly summarizes his argument at the end of each chapter. In essence, Wynne-Edwards' thesis is as simple and universal as Darwin's. Animal populations are regulated, not directly by food supply, but by intraspecific competition involving behavioural interactions for territory, social status, or other conventional prizes, in which some are victors and some are losers. The hypothesis has not been without critics. On the one hand, population geneticists have attacked it on the grounds that any adaptations by which individuals accept restrictions on their own fitness, except possibly in favour of close relatives, could not evolve by natural selection. On the other hand, the idea that competition for territory and social status is part of our animal heritage, a view popularized by Mr Robert Ardrey and others, has appeared distasteful to some liberal sociologists. Wynne-Edwards now appears to accept the first criticism: 'The general concensus of theoretical biologists at present is that credible models cannot be devised by which the slow march of group selection could overtake the much faster spread of selfish genes that bring gains in individual fitness. I therefore accept this opinion.' On the second criticism he makes no concession: 'Though it may come as a surprise to some people to find such emphasis laid on the competitive nature of society, not much reflection is required to see that the cap fits. Do we not find in our own social world, the leaders at the top, the winning teams, the houseproud, the status-seekers, the coveters of pay differentials, of rank, titles and honours, which reveal the persuasiveness of sophisticated, and that is to say social, competition in our lives.' To the convenors of this symposium, these criticisms seem in any case to be tangential to the discussion. The issue is whether or not the postulated social mechanisms exist. One cannot dismiss them on the grounds that their evolution is inconceivable or that their existence is socially inconvenient. For these reasons, while controversial discussion of selection mechanisms or indeed of social implications will be greatly welcomed, these topics do not, in fact, form part of the invited papers.

It remains for me to thank all the contributors who have produced the excellent papers which you hold in your hands, and to thank you all for coming. Finally, the most pleasant and important duty of all is to welcome and introduce Professor V. C. Wynne-Edwards to our symposium.

Discussion of papers by V. C. Wynne-Edwards, J. R. Krebs and C. M. Perrins, and J. R. Flowerdew

JEWELL: The red grouse story is a compelling example of social control, but in considering the several examples given—man and overfishing, myxomatosis, laboratory populations—we must be cautious because all are man-made situations, even heather moorland. Direct limitation of density by food availability has been demonstrated for many large herbivores: African buffalo, wildebeeste, Soay sheep. But here the food resource—pasture—is very resilient and recovers within a season from heavy use: this is in contrast to the long recovery time required by trees.

WYNNE-EDWARDS: There is a quick answer to that: the red grouse has a close relative, the ptarmigan, which lives in an unaltered arctic-alpine environment above the heather zone in the same area; it is an interesting parallel to the red grouse, because it does almost everything the same. Below 3000 feet we also have plenty of natural unmanaged heather inhabited by red grouse, which operate their normal control system. I myself wonder whether the herbivores in East Africa could not damage their grazing; they certainly can in arid regions, and in some places desertification has resulted from overgrazing by livestock of primitive people. It may be that there is some grazing which cannot be damaged but I think that needs to be demonstrated. It is very clear that pasture can be damaged by livestock in arable and enclosed land in almost every part of the world.

VAN DE VEEN: Fishing in the North Sea is not a good example of control of fish by predators, but an almost perfect example of the regulation of predator population by the food resource. Nowadays commercial fish are kept at very low levels by a greatly increased energy input by a strongly reduced stock of fishermen.

WYNNE-EDWARDS: My point is that you can take a crop of a certain size, called by the fishing industry the 'maximum sustainable yield'. As long as you do not exceed the maximum sustainable yeild it will go on forever. But if you put more energy into the system and you exceed that, you will drive the stock down, and that is what has happened, and what always happens when you put too much energy into wildlife exploitation. If you have a pair of wings, you can do it too, or if you run on four feet, or whatever it is, you can have too many predators and then you will reduce the stock and the yield.

CLARKE: Is it feasible, Dr Krebs, to rear great tits in the laboratory to see whether those differences which you have suggested are, dare I say, inherited? Or are they something which is imposed on them by their early environment?

KREBS: The short answer is that we don't know whether the differences are environmental or genetic.

CLARKE: But can you experiment in the laboratory with great tits? Can you take them out of their natural situation and see whether they are better or worse than others?

KREBS: We could do that. An alternative would be to look at the heritability of various components of survival in the field, which people are doing in Holland, in particular.

PERRINS: Since body size is known to be heritable, and body

size influences survival, that partly answers Dr Clarke's question.

JELLIS: I have reared and observed many great tit broods in 'semi-laboratory' conditions and one point of interest is the differential feeding of the young in the nest. In a poor year many fewer will survive to fledge, but there is the alternative that, say, 6 out of 8 will fledge but 2 will be lighter. This may, as you say, be heritable, but on observation it looks far more likely that it is environmental, that there has been less food to go round and therefore some are lighter when they fledge. Would you agree with that, Dr Krebs?

KREBS: Yes. I think it is hard to disentangle cause and effect, because one then asks 'Which young got fed more, was it the ones which were a little bit bigger to start with?'

CLARKE: Dr Flowerdew, what determines the length of the breeding season—the earliness or lateness of the onset of breeding in wood mice, and for that matter in red grouse, great tits, and other animals?

FLOWERDEW: In the wood mouse the length of the breeding season is correlated with the winter food supply, so it is likely to go on longer if there is a large amount of food available in the autumn. This is not to say that food is necessarily the major cue which initiates breeding, which I think is basically photoperiodic, though this can be modified by food supply, perhaps also by temperature, and maybe also by social factors, particularly towards the end of the breeding season. There are examples where in high population densities bank voles are found to end breeding earlier than normal.

SWANSON: Dr Flowerdew, what causes the decline in overwintering wood mice, since they are the largest and most dominant animal?

FLOWERDEW: We know that some of them move out of the study areas, so dispersal is one cause of decline in the population of wood mice from winter to spring. I think, also, there are times when food supply is very short so there may be starvation. And, as I have indicated, predation will take a proportion, although this is something we know little about. Overall, because there is a very good correlation between increase in aggression and the start of breeding, I think that as breeding starts, adult males and probably adult females are competing with one another—some are dispersing and some are staying behind, and this is probably the major mechanism for determining numbers in the following summer, although it is obviously related to the food supply available. If there is a lot of food available in the winter, it is likely that there will be a high density; individuals can live during the spring and carry on into the summer at a relatively high density. If food supply in the winter is not great, then numbers will decline and probably behavioural interactions will sort out which ones are going to remain and which ones disperse. What happens to the dispersed animals nobody knows.

WALLIS: Referring to food availability in spring, you said that you detected a reduction in the amount of overwintering seeds, and at this time when food is reduced you recorded a higher incidence of insects in the habitat. If you look at the stomach contents of spring-caught animals you find a very high proportion of arthropods—what do you feel is the significance of this?

FLOWERDEW: In the one year that I did look at this there was very little insect material. Two or three years earlier, Watts did find a lot of winter moth larvae in the stomachs. There were more in the spring, towards March and April, than in the winter; I am not sure when the winter moth larvae drop. Obviously insects can form an important part of the diet of the woodmouse, more in the summer and late spring than in the winter.

WALLIS: You say it is late spring—well, this is the 'bottleneck' period for supplies of overwintering seeds.

FLOWERDEW: Yes, the bottleneck is, I would say, around February–March. Numbers do decline after this but I feel that food shortage at this time rather than later is important.

GREENWOOD: Could I raise a point about the great tit? The factors that were identified as influencing the population were all acting intraspecifically. Recently Dhondt (1977*) seems to have shown in a continental population that the density of blue tits is having a density-dependent effect on the breeding success of great tits. At high combined densities of blue tits and great tits, blue tits are doing proportionately better than great tits,

* Dhondt, A. A. (1977) Interspecific competition between great and blue tit. *Nature*, **268**, 521–523.

and at low densities great tits are doing proportionately better than blue tits. Is there any evidence that this occurs in the Oxford great tit study? If it does, and is incorporated into the model, would it produce a better picture?

PERRINS: We have tried to analyse the blue and great tit data in the same way as Dhondt and we do not get his correlation, at least significantly, and probably not at all. If there is a difference there is only a very minor one. There are, as Dr Krebs said earlier, some big differences between continental and British situations, so this is not meant as a criticism of his data; but we cannot reproduce them.

On the national scale, if you look at ringing records, the total number of birds that are reported every year to the British Trust for Ornithology as being ringed each year as nestlings, over the last 6 or 7 years, you find the numbers of great tits have declined quite markedly, and the number of blue tits has gone up. I think this is a genuine shift in population numbers in both, and it does seem to be reciprocal. This may suggest that there is some form of interaction which we have not succeeded in identifying. I don't know!

Session Two

Chairman: V. C. WYNNE-EDWARDS

Chairman's introduction

You come to a symposium like this hoping to learn something new and useful for your own work, and when you take part in the presentations and discussions, you may also find people who can show you where you have gone wrong. These are valuable things, and I am appreciating them myself today.

The first speaker this afternoon is Dr Alison Jolly, who is on the staff of the University of Sussex. She is a first-degree graduate of Cornell and a PhD of Yale. Her fieldwork was done on Madagascan lemurs. I notice in the references at the end of her paper that she has already written two books, one on lemur behaviour and the other on the evolution of primate behaviour, and she tells me that she is now working on a third.

Dr Michael Stoddart, author of the paper of 'The place of odours in population processes' was originally an Aberdeen graduate, although he has been for a number of years at King's College, London. He is one of the people responsible for getting this symposium off the ground. I think we are all grateful to Professor Ebling in particular, on whom a very big share of the organization has fallen, and to the Institute of Biology for setting it up.

Discussion of papers by A. Jolly, J. Deag, and D. M. Stoddart

WALLIS: A lot of terms which scientists use, particularly in behavioural studies, are not defined at the outset. It is assumed, for example, that everyone knows what 'play' is and what 'dominance' is, but these things are rarely defined and it is this which I feel leads to much of the confusion when comparison between studies is made.

DEAG: I agree that one has to be very careful to define terms. Although I may not have defined some terms, e.g. 'fitness', when presenting this paper, I have done so in the written version. Incidentally, with reference to the problems of estimating fitness, I should have referred to the discussion in R. C. Lewontin's book *The Genetic Basis of Evolutionary Change* (Columbia University Press, New York). In this he concludes (pp. 232–239) that: 'To the present moment no one has succeeded in measuring with any accuracy the net fitnesses of genotypes for any locus in any species in any environment in nature.'

WYNNE-EDWARDS: I would like to ask Dr Stoddart whether you can distinguish between odours becoming increased by the owner's dominance in the hierarchy, and dominance becoming established through odours?

STODDART: It would be nice to be able to dissociate these events. What studies have been done on rodents suggest that dominance is the cause and changed odour is the effect. What we don't know anything at all about is the nature of the biochemical change, how the same food base and the same body chemistry can create a new odour with an obviously different message.

GEIST: You may be aware of the work of Müller-Schwarze on the black-tailed deer of North America in which some of the things that you stated are nicely reflected. For instance, the juvenile males still smell like females and are able therefore to cover up their maleness in the neighbourhood of a large rutting male.

I wanted to bring something else to your attention: you stated in the example of the red deer that you did not know of any obvious odour differences between the large male and the small male, but there is one very noticeable one, namely the frequency of urination against the belly varies enormously between young and old stags. Presumably the intensity of olfactory signals alters with age. The same principle operates in the domestic goat (Schank, 1972, Zeitschrift für Tierpsychologie) in which the frequency of urine marking varies with age. The urine stream is aimed at the beard. I have had to autopsy a large number of male goats, and I can assure you that the difference in smell is *extremely* great, particularly on a warm day.

STODDART: I'm sure that is so. Could I ask you whether these differences in odour occur in the bladder urine, or are they a result of bacterial decay once the urine is outside the bladder?

GEIST: We do not know. We have a clearer picture of what happens in the black tailed deer, which produces lactones. Müller-Schwarze suggested that these lactones are ingested with the food, excreted with the urine, and selectively absorbed on the long tarsal gland hairs. He took urine samples, while the males urinated, from both above and below the tarsal glands, and discovered that there was a deficiency in the lactones in the samples taken below. The olfactory picture is complex, since there are some 26 lactones, but it is of interest that males and females differ in only two of them.

STODDART: One must be aware of the possibility of substances being produced by bacterial decay, and I purposely did not mention this because I think it raises many difficult evolutionary problems. As far as the red deer, and I suppose also the wapiti is concerned, there is the further interaction with substances from soil in which the animal has been wallowing, and the total odorous message is quite unknown.

SWANSON: I want to ask why you felt such a strong objection to using the word 'pheromones', for instance in the Bruce effect in mice, where the suppression of pregnancy appears to be a very specific response.

STODDART: The Bruce effect is capable of being modified, depending on the amount of time that has passed from fertilization; it will only occur within two days of fertilization. But its action is exactly analogous to that of a sex lure in the gipsy moth or any other insect example. Perhaps my cautionary words were a bit too strong, and this is a situation in which the term 'pheromone' could be acceptable. The Whitten effect, i.e. the synchronization of oestrus, may be another. In these situations, the odours are functioning exactly like pheromones—in fact they are pheromones, but I think the danger is that when you can use the term for those two effects you also tend to use it for other ones of a different nature.

SWANSON: I wanted to mention some work that we have done with gerbils, which have very large scent glands, usually much bigger in the males than in the females. We have studied gerbils in free breeding populations in enclosures, and we find that if you leave the mother and the father in the enclosure, of all the females present the mother is the only one who continues breeding. It is interesting that in these circumstances the scent glands of the female offspring do not develop so that these females become mature in age, but their reproductive system does not mature. In male offspring we get a parallel effect; the scent gland appears but it does not grow as large as in the father and, particularly, the marking behaviour of the sons is very much suppressed as long as the father is present. If a young male and a young female gerbil from an enclosure are taken and put together in a cage, then the scent glands immediately start to develop in the females. In the males, similarly, the scent glands increase in size and scent marking is initiated.

STODDART: Are you certain that in your experimental technique you are not suppressing the development of sex hormones through, perhaps, a dietary deficiency or any other factor, and that this accounts for a failure of the growth of the scent glands? And are you sure that what you are dealing with is the odour of the other individuals rather than some other aspect of their presence?

SWANSON: We are investigating this at present. I don't think the explanation is likely to be nutritional because there is plenty of food. But we don't know the nature of the cue produced by the mother or even whether she has to have a scent gland. The first experiment we are going to do is to remove the scent gland. As we don't know whether it matters if she is reproductively active, we are going to ovariectomize her—it may be just the presence of a large animal which is inhibiting the offspring. We shall try to find out by what mechanisms the mother influences the daughters and the father the sons.

STODDART: Mitchell showed ten years or more ago that the presence of the ventral gland is not necessary for reproduction in the Mongolian gerbil.

SWANSON: No, I don't think it is, and nor is it necessary for marking; animals continue to mark when the gland is removed.

BLAKE: Odour has been discussed as a means of communication among primates and other animals. In man one hears numerous anecdotes concerning smells of people of different races. Has any study been done on racial odours or odours in reproductive behaviour in man?

STODDART: There has been very little work that one could refer to as objective study. There are plenty of anecdotes, but I don't think this is the place for anecdotes.

EBLING: There are certainly plenty of anecdotes; Havelock Ellis, in his *Psychology of Sex*, Vol. IV, published in 1905, recounts a number of very entertaining ones. But I could offer a serious comment on human racial differences, which perhaps is related to Dr Stoddart's observation that odour in deer is related to the level of male hormone. In man, there appear to be distinct differences in odour production and odour perception between caucasoids and mongoloids. Chinese and Japanese are said to be conscious of the stronger smell of Europeans, and even to be able to distinguish between individual odours. They also have much less body hair than Europeans, and it is interesting that both body hair and the skin glands which produce odour are androgen-sensitive structures. Whether the racial difference is actually related to testosterone levels is doubtful; it is probably a difference in the response of the skin.

WYNNE-EDWARDS: Dr Deag, you said that primates were in many ways most unsuitable animals for studying population processes because of their long generation time, but I think they are very good in other ways, being diurnal and visible and to some extent individually identifiable without marking. I wonder whether either you or Dr Jolly would like to say anything about the ease of identification of primates, and whether it gets easier to identify them the more they resemble the hominidae.

JOLLY: I argued that they are lovely animals, not only because you can easily identify them (though it would still be a great help to have more tattooed or tagged populations that could be passed on easily from one observer to another), but because so many of the behavioural interactions are obvious. It is not hard to work out dominance hierarchies. Even in lemurs which have a large repertoire of scent communications, all the marking is visible, though we may have missed a lot. There are double gestures, in which you can photograph the animal doing something, and then go and pick up the scent it has left when it has done it; lemurs clearly react both to the visual and odorous components of the signal. The difficulties lie in the length of time it takes to get population data and the problem of finding a population in anything like an undisturbed habitat.

DEAG: One of the most interesting things about macaques and baboons is that they live promiscuously in multi-male groups. As a consequence there is a paternity problem—they don't know who are their fathers and paternal sibs, and the fathers don't know who are their offspring. If you want to follow the reproductive success of males then this creates major problems. With monogamous primates, such as gibbons, the problems are less severe and, apart from the animals' longevity, are similar to those experienced in studies of red grouse and great tits.

JOLLY: The problem is not just time and complexity of paternity, but also space. Referring to migrants, we know very little about where an animal that leaves a primate troop goes. Ones that have been observed rejoining troops always join groups near the ones they left, because the observing primatologist or his student was in the same area. Picking up a macaque even 25 kilometres away, like the great tits that get returned ringed, is very difficult. The only work I know on colonization of a new area by primates is that by Kavanagh on a group of Colobus who have colonized a poor farm area in West Africa. Their behaviour has changed from that in the wild. For example, they do not respond to a man with loud alarm calls, but with a very quiet click, which the farmers have not yet learned to recognize, and they treat dogs as people, with quiet calls, not as jackals or hyaenas which are predators to be shouted at.

GEIST: May I bring up one point about large mammals. It is true that they can be observed very well, and precisely for this reason, and because we can often see the causes and effects of their actvities, we can also make a few statements about population control mechanisms. Professor Wynne-Edwards has proposed that social behaviour probably arose as a means of controlling resources. Yet the mammals that I am familiar with, mainly ungulates from the Northern areas, show an insignificant amount of behaviour that can be related to defence of resources. This is quite understandable. If you ask the question 'What must an individual do in order to maximize its reproductive fitness?' a number of deductions follow. The individual must maintain access not only to scarce resources but also to the most desirable mate (which means that there must be social behaviours which will attract, hold, and maintain a mate that is its equal or superior in enhancing its reproductive fitness). There is another rather ugly deduction, and that is that it must interfere with others in such a fashion as to reduce their reproductive fitness. So there are at least *three* causes of simultaneously operating social behaviour. Mr Chairman, we suspect that you were right ultimately, that social behaviour may indeed function to limit the reproductive output of the population, but it does so in a way very different from anything discussed here today. I hope I may be permitted to elaborate. Large ruminants are animals which ingest very coarse forage, forage which is widely scattered and almost pointless to defend. There are, in fact, very few known examples of these animals defending forage. However, there is another way of getting at the competitor, and that is by raising its expenditure of energy and nutrients for maintenance. If there is any way to increase an individual's maintenance costs then its reproductive output can be influenced. We are at present working on the hypothesis that this can be influenced by very subtle forms of behaviour that cause excitation and raise the cost of maintenance. Excitation can cost anything between a few per cent and almost 250 per cent of the requirement for maintenance, and that is quite considerable, particularly bearing in mind that an individual can only ingest, at the most, about twice maintenance. If you were in a position, as a dominant, to cause excitation in a subordinate, then you could reduce its reproductive output as the quality of forage became poorer.

SWANSON: I have some information relevant to Dr Geist's remarks about energy cost in population control. We compared hamsters and gerbils living in similar enclosures. Although they both controlled their population to about the same size, and we never ended up with more than about 15–25 animals in an enclosure, the mechanisms appeared strikingly different. In hamsters the females were pregnant all the time. They continued to mate and kept having litters, even though none of the young survived. If one litter did survive to maturity, the young females of the first generation also started breeding and were always pregnant. Their litters did not survive either, which seems a very costly way of limiting population. In gerbils the process was completely different. If the mother and father are present, none of the female offspring become pregnant. One suggested reason for this difference is that, in the wild, hamsters are solitary, whilst gerbils live in colonies. Hamsters have not needed to evolve the type of controls used by gerbils.

WYNNE-EDWARDS: Do hamsters eat their dead young?

SWANSON: Sometimes they eat them, and sometimes they don't. They are certainly not short of food. It has been suggested that it is not usually the mother who kills the babies, but other individuals in the enclosure. Cannibalism may perhaps be connected with unfamiliarity; mothers do not usually meet babies that are not their own, so they may consider them as a source of food and treat them as such.

WYNNE-EDWARDS: I would like to return to the business of recognition. I am fascinated by our human capacity for recognizing people. I said to someone only today that when you get to my age, people's names just fly away from you, but their faces do not. There is very little decline in one's memory of faces. As you get older you seem to know more faces. The memory store for these complex physical features seems to be relatively inexhaustible in comparison with the memory for personal names. I was wondering whether you recognize individual primates by facial features, in the way we recognize our fellow men, or is it all by having nicks in the ear, etc?

DEAG: I think that many primatologists do recognize individuals by facial features. You may at first recognize an animal because it limps, has a piece of fur missing, or a nick in the ear, but once you have spent some time with a group you notice that their faces are different. There is a good series of chimpanzee faces pictured in Jane Van Lawick-Goodall's paper in *Animal Behaviour Monographs* Vol. 1. You have only to glance at them to see how much variability there is in lip shape, ear shape, amounts of bare skin on the face, or of missing hair, size of cheekbones, and so on.

CLARKE: It surprises me slightly, as exclusively a laboratory worker, that no one here who works in the field has taken up Dr Deag's rather provocative remarks concerning the nature of science in the field. I wonder whether anyone would like to comment on his recipe for the construction of science based on the ideas of Popper, a development which depends really on the state of the subject. It seems that some of those theories, created in the way Dr Deag has suggested, are not themselves testable. I would have thought that this might provoke a response from field workers whose difficulties are very considerable compared with ours in the laboratory.

DEAG: Yes, I acknowledge that it is a lot easier to use this approach in the laboratory and this has in fact been done for a long time. Platt (in the paper referred to in the references to my paper) showed how the approach has been successfully applied in molecular biology. You have, however, only to look at the work in Oxford on gulls and great tits and in Aberdeen on red grouse, to see that the approach can be applied in field work. The work will be much harder with long-lived, promiscuous animals such as the anthropoid primates.

ZAHAVI: Since technique has been attacked, I would like to suggest that people in the field are speculating too little. By this they handicap their ability to see. I think that speculations should come before the start of field work. The currently accepted philosophy of the natural history sciences allows speculation only one step ahead before demanding quantitative data. This diminishes the chance of finding new qualitative facts. If one speculates three or four paces ahead, one has a much greater chance of finding new qualitative data.

DEAG: Dr Zahavi is well known for his interesting speculations—and I would not wish to dampen his enthusiasm in any way. What is important is to *know* when you are speculating, and not to use the term 'explanation' with reference to speculative theories. So let us speculate by all means, but not fool ourselves into thinking that we have found explanations, when all we have are hypotheses that are simply speculative interpretations of the data.

COHEN: I would hate us to finish this session with such a naive view of the philosophy of science as we have just heard. A lot has happened since Popper! Particularly Lakatos. I think that people who are concerned in this area are no longer using the words 'explanation' and 'research programme' in quite the same way as Dr Deag. I don't want to go into this, because I believe I understand it imperfectly, but I hoped there might be a philosopher of science in the audience who would take issue with Dr Deag. As there was not, could I recommend people here who may not be concerned to find 'explanations' but would like to think of research programmes as a continuing series of questions (as Dr Zahavi was suggesting) to read some Lakatos*.

DEAG: I do not pretend to be fully familiar with the philosophy of science! Those of you who listened carefully will have noticed that I hesitated somewhat when I used the word 'explanation', for you have only to read Popper to realize how careful you have to be in using this word. Its use has clouded the issue. The essential point is to appreciate the difference between imagining you have evidence to account for a biological situation or event and actually producing documentation.

JOLLY: I think that a lot of us pick the hypothesis first and then work back and try to find explanations which fit what we

* e.g. Lakatos, I. (1970) Falsification and the methodology of scientific research programmes. In *Criticism and the Growth of Knowledge*, ed. Lakatos and Musgrave. pp. 91–196. Cambridge: University Press.

believe. I think that theory has now got to the point where it is almost too good; we already have Darwinian and neo-Darwinian evolution and now we have got Sociobiology and the Selfish Gene, too, we can explain anything. Is not all we are saying, in various ways, is that we should be cautious? Is the thunder going to strike so that we will have to start with a completely different set of theories, or can we really fit everything into the ones we have already got?

DEAG: A major point made by Popper is to emphasize deductive reasoning; he has little place for induction. I do not personally feel qualified to come down on one side of the fence or the other. But I can, however, see with my own eyes that inductive reasoning has not led us to the goal which some people claim. That is why I suggested that we should pay more attention to deductive argument.

EBLING: Perhaps I could provoke Dr Deag by asking him how he would apply some of his criticisms to the writings of Darwin. However, what I really want to ask is where this leaves the hypothesis we are discussing. Is some sort of summary possible at the end of our first day? Wynne-Edwards originally presented his view using much the same approach as Darwin in the *Origin of Species*, by assembling all the evidence. How much experimental proof has been obtained since then, and what have we heard new today? We have had material about the red grouse from Professor Wynne-Edwards himself, and we have heard facts about the great tit and the wood mouse, some of which appear to support Wynne-Edwards' views and some of which might lead to different conclusions. The primatologists have stalled on the grounds that there isn't sufficient evidence to say anything, and this has led us into a philosophical discussion on what the nature of that evidence ought to be. Perhaps Professor Wynne-Edwards himself is going to tell us about what he thinks of the position at this moment?

WYNNE-EDWARDS: I can say that I am content that my hypothesis should still be alive after 18 years since it was first printed as a paper in the *Ibis*. It is not dead yet by a long chalk. It takes time to sort these things out, and however much effort is put into organizing a meeting of this kind, I don't think you can expect it to solve a major controversy. It serves to interest more people in the subject and its problems, and to see how their work can contribute to it. It also makes them question their own objections and alternative explanations. I feel we have had an entertaining and interesting day with good speakers, and I would like to thank them all very much indeed.

Session Three

Chairman: R. D. MARTIN

Chairman's introduction

The first paper is on the question of stress, and I thought it might be appropriate to say a little about Hans Selye, who did so much to lay the foundations of work on stress, and in fact gave the word 'stress' to several languages. For example, in the French and German languages the word 'stress' apparently did not exist until his work became known. He eventually ended up with an institute with dozens of people working on stress and producing an annual bulletin containing the latest results. The reason I particularly want to mention him is that the way he started out on his research is rather interesting. As a medical student he took part in an examination involving inspection of patients with the aim of identifying specific disease conditions. Of course, the students involved were being led to recognize individual symptoms of particular diseases, but the thing that struck him was that all of the patients had common symptoms *because they were ill*, that in fact there was a syndrome of characteristics shown by all of them because there was something wrong with the organism as a whole. And this, he claims in his autobiography, is what started him off as a medical student thinking about stress and eventually led to a lifetime's research. It is a nice story, I believe, because it illustrates the way in which an imaginative biologist, even in a situation in which everything seems to be cut and dried, can come up with a new hypothesis to explain facts which other people have not hitherto put together. It gives me particular pleasure therefore to introduce our first speaker, Dr Clarke from Oxford, because in fact I can trace some of my own research to a series of lectures he gave on reproductive biology quite some time ago. I particularly remember him as giving clear-cut lectures which were also peppered with a nice sense of the amusing. One thing in particular I remember is his explanation of the human menopause by saying that when God created man, he did not expect him to live for more than 30 years. I hope that when he talks to us about stress he will have some equally penetrating things to say.

Dr Mackintosh continues the basic theme of the relationship between detailed studies in the laboratory and what is going on in the wild. He has been working for 20 years on his particular species, the mouse, and moving progressively towards a synthesis.

Discussion of papers by J. R. Clarke, J. H. Mackintosh, and L. R. and R. A. J. Taylor

MARTIN: Dr Clarke has given a detailed, clear account of exactly how the stress mechanisms operate in so far as we know at present. Still unresolved is the question of how far that mechanism may operate under natural conditions to limit populations or in other ways. Certainly in captivity there seems to be little doubt that reproduction is adversely affected by the phenomenon that we can term 'stress'.

WYNNE-EDWARDS: Dr Mackintosh, you said that it was surprising that at lower densities there was more aggression in the cages. In fact, the same thing happens in the wild; for instance, when red grouse are at low densities and territories are large, there is much aggression. More effort is needed to keep a large territory. Is this not a general phenomenon?

MACKINTOSH: In several of the papers I have looked through recently the authors have said: 'Surprisingly in this case increased density did not produce increased aggression', so the expectation seems to be there.

SWANSON: I think it might be relevant to mention a few of our findings with gerbils in which we also set up populations in enclosures, although ours were much smaller than those of Dr

Mackintosh. When the mother and father were with a family in an enclosure or a cage the parents inhibited reproduction in young animals. It was interesting that, although the mother kept producing litters, the population stabilized at around 17–25 animals, in the enclosures or cages alike, even though there was a lot more room in the enclosures. In the enclosures stabilization was achieved by two mechanisms. First, although the mother continued to have litters they did not survive—this seemed to be due to neglect rather than cannibalism. Second, when we set up populations in enclosures of young animals without parents, they matured much later than did controls living in cages. In fact, the inhibition was similar to that observed previously in the presence of their parents. Until the age of about 20 weeks the animals remained sexually immature. They lived peacefully without fighting and there was no evidence of an hierarchic structure. At about 20 weeks there was a social upheaval—a lot of fighting occurred, both males and females started to mature and develop scent glands, and in due course one female emerged as a breeding female and one male emerged as dominant and was the only one to show scent marking. In contrast, similar groups of litter mates living in cages matured at about 12 weeks. All females started breeding with little conflict and without any sort of hierarchy being established. Thus not only the presence of the parents but also the size of the enclosure had a marked effect on the social organization.

CLARKE: I would like to ask the Taylors to what extent their work depends on an understanding of the physiology of aphids? It seems to me that what they have demonstrated, in a very persuasive fashion, is how far one can go in analysing the population ecology of a species, but they have made no reference at all to the sorts of things that I discussed.

TAYLOR: We do not speak as physiologists but have tried in the written paper to summarize some of the work that has been done on the physiology and behaviour of aphids and here we are very much indebted to Professors J. S. Kennedy and A. D. Lees who unfortunately are not present. The mechanisms have been very well investigated. The initial stimulus is largely tactile, that is to say, if you touch an aphid in the right kind of way at the right time you can turn it into a migrant, or at least you can turn its children into migrants. There are other environmental, as well as genetic, components but I would like to emphasize that it is often the offspring and even the grandchildren which respond and there can be time delays which make the migratory response complex and difficult to detect unless it is deliberately sought.

Although the actual physiological pathways are now quite well known, they are used only to illustrate the model; it was not constructed from them.

COHEN: May I make two separate points? The first is a plea to recognize that the words 'reproduction', 'multiplication', and 'breeding' have somewhat different meanings. I think that on several occasions you used 'reproduction' to mean breeding or multiplying. I was unhappy about this, when the population that is produced is different from the one that you start with.

The second point: Dr Russell Coope and his co-workers studying Quarternary fossil beetles, have shown marked discontinuities in the specific composition of historical sequences of beetle faunas, that are best explained as the result of sudden and drastic changes in the climate rather than as the result of any evolutionary changes. This is because the species concerned are still living today, even though some of them have modern geographical ranges remote from Britain. For example, the commonest large dung beetle in England during the last glaciation is now confined to the high plateau of Tibet*. This picture, then, is one of enormous changes in distribution on a rather larger time scale than you have used in your model, but both illustrate the transience of insect distributional patterns.

TAYLOR: To incorporate the work of Coope and other palaeontologists we did put 'evolution' in the original title, but it was taken out to save space. Many of these insect species have existed for incredible lengths of time and discussions about changes of environment during evolution usually emphasize temporal change. There is however remarkably little that has happened to the environment even over very long periods that cannot be equalled by simply walking to another place on the planet as it exists now. Although the maps we showed relate to this small island (and that was the biggest experimental arena that we could have managed) I am quite sure that the argument

* Coope, G. R. (1973) *Nature*, **245**, 335–336, and Coope, G. R. (1967) In *Insect Diversity*, Symp. Roy. Ent. Soc.

applies over longer periods of time and larger areas. By finding the right place, or in fact by 'evolving' into another place, an organism is doing exactly the same thing, it is practising, if you like, for changes that will occur through time.

JOLLY: Dr Taylor, could you enlarge on the biological significance of your final graph on population variance and density? In the graph of the corn earworm the variation looks like a normal bell shaped curve. Is this the same for all species? I can imagine, for instance, the population of the United States turning into a U-shaped curve, with no one in the small towns, virtually everyone piled on top of each other in Manhattan, and just one person left at the other end of the curve studying mountain sheep in the Rockies!

TAYLOR: This is obviously one of the controversial areas. Yes, we all do behave in the same way. These distribution patterns are common to all organisms although they do not actually become U-shaped. When the human population increases, it doesn't increase uniformly all over the country—people actually move into the cities, increasingly so, and denude the countryside.

That is happening in most of the countries of the world now where mean population density has become so high. It is a process which involves not just increase by reproduction or multiplication, but also by movement.

WYNNE-EDWARDS: You have a series of density maps comparing abundance in different years for some of these moths. Does your system of collecting data allow you to distinguish between annual differences of local production in a population more or less permanently located in Herefordshire, say, and differences caused by mass movement from one point to another? One is familiar with the fact that some insects have very big years and then they decline again, and as you have shown they don't necessarily do this all over the country at the same time. One also knows that there are some Lepidoptera, for instance, that are highly localized: they are like rare plants, you have to go to the right place to find them, and in some years there are fewer than in others. What I cannot get clear is how many of the variations in your maps are due to what I would call autochthonous variation of populations, which undergo ups and downs in limited areas, and how many to actual bodily movement across the country, so that you have a shifting population without a permanent centre which shows different densities at different times in different places?

TAYLOR: We do not detect movement simultaneously with the measurement of population density. This is one of the most difficult things to do, as I am sure everyone knows. The argument is based on a collection of pieces of evidence; that organisms do move, that they move in response to density, that the resulting population distributions differ and that there is a strong behavioural component in those differences. The inference is then that density-dependent movement is at least a small component of the resulting distribution. The model says it does not have to be a large component; it does not say that multiplication is unimportant, but that stabilizing feedback systems can depend on density-dependent behaviour, and the distances involved do not have to be very great. So there is no implication that, when those patches move around on the map, all the same individuals are moving from one patch to the next but, equally, most populations have no centre. The implication is that individuals are constantly searching, competing, or avoiding competition; a 'fight or flight syndrome'. They are not compelled to stay and fight; they can always avoid the issue. Dr Jolly's remark fits the situation so well that I cannot improve on it: 'the one migrates that has most to gain'. It doesn't have to be the one that is a total failure because, if the environment is going to change, which all environments do, then everyone has something to gain by getting out before this happens. The anticipation of change is precisely what aphids have 'learned' to excel in throughout evolution because it isn't wise to stay too long in the same place.

JEWELL: Could we throw the discussion open a little more widely? The last paper was stimulating in showing how an approach that has nothing to do with what individuals do and how they behave can reveal aspects of population control. On the other hand, Dr Jolly, in her paper, was forced to say that one can see so few individuals in studying a primate population that one cannot get much information about population densities, extent of movement, and so on. Perhaps these two approaches must eventually come together to give us the information we want. Ironically, it may well prove that studies of human behaviour in ethological terms, playing back on to people looking at primates, might give a way of discerning certain aspects of their behaviour that as yet have not been recognized as agonistic, and are causing them to move or reduce their density. Now this (and I felt it related to some of the other papers this morning) is one of the problems of using small mammals as models; most of the time you simply cannot see them. Instead, you trap them and imprison them. I feel that after long years of study of small mammals, this work has ground to a halt for this reason. To confuse multiplication and immigration and emigration in these studies raised yet another difficulty, and I thought that the best way to try to control it was to use enclosed populations. Accordingly I set up some enclosures to look at bank voles. I also thought one ought to try to keep numbers, and therefore density, constant whilst changing the aggressive interactions of the individuals, and to do that I thought that if one castrated a large number of the males in one enclosure and compared them with another population of exactly the same numbers, one might get some clue as to whether these factors were important. Mr Gipps, who is working with me, will say something about our results.

GIPPS: There were two large field enclosures, each 550 square metres in area, and into each enclosure I introduced similar populations of voles, 25 males and 20 females. Twenty of the males in one enclosure were castrated. I showed, in a neutral arena study, that the castrated males were less aggressive than intact males.

The experiment lasted 10 months, and in both enclosures there were great increases in numbers, so that at the peak there were 180 animals in 550 m^2, about 3500 voles per hectare.

In common with many previous studies on mice, there was very marked reproductive inhibition in both the animals who were introduced into the enclosures and those who were born there, to the extent that none of the animals born in the enclosures ever bred at all, and very few came into breeding condition; however, there was a marked difference between both the final numbers and the rates of increase in the two populations; the numbers in the enclosure containing a large proportion of castrated, non-aggressive males, rose significantly faster and to a higher level than did the numbers in the enclosure containing the larger number of aggressive intact males. I marked neonatal animals when I caught them in traps and in nest-boxes and I found that a significantly smaller proportion of the neonates were reaching the trappable population in the enclosure containing the intact adult males. I therefore concluded that the differences in rates of growth in numbers in the two enclosures, and in final numbers, were due to mortality between birth and first capture, and this again has been shown in previous studies of mice in large enclosures.

There are two prime problems with the approach: one is that I have no idea what effect manipulation of other age and sex classes might have had—I was dealing simply with adult males. Secondly, the densities reached in the enclosures were so high that the experiment might be criticised on the grounds that, although the conditions were field conditions and they were outdoor enclosures, the results might not really bear much relationship to the real world of bank voles.

Recently Dr Lidicker* put mice in field enclosures of 385 m^2, slightly smaller than mine, and obtained a final density of 60,000 per hectare, compared with my 3500. This says something about the differences between mice and voles, but whether one can draw general conclusions is questionable.

MARTIN: I think that comment brings out the fact that one really has to work towards a marriage between laboratory and field studies.

ZAHAVI: I have been studying a population of birds (Arabian Babblers) for six years. They are living in groups, and we can identify individuals, and follow their dispersion. Sometimes individuals disperse because they are forced by the behaviour of their dominants, sometimes of their own choice, and you can follow them as they monitor the available space to make a decision. When you record all they do, they make very wise decisions. Another feature of these birds is that they have so-called 'helpers' at the nest. Our data show that if there is any adaptation in these group-living animals for such activity, it is to control group size rather than to help. Most birds and mammals reproduce very well without any help. It is interesting that 'helpers' are common among group-living animals with

* Lidicker, W. Z. (1976) Social behaviour and density regulation in house mice living in large enclosures. *Journal of Animal Ecology*, **45**, 677–697.

territories, when it is in the interest of many individuals, especially non-breeding ones, to control group size. People who do not have population data about the groups, but just watch them, call the birds 'helpers', but as they often kill their kin 'controllers' might be more appropriate.

CLARKE: I still have a feeling of unease relating to the question I have already put to Dr Taylor. What role does the understanding of individuals play in explaining population phenomena? Apparently one can go a very long way in describing and even predicting population phenomena without feeding any physiological data into the model. Is it important to find out what individuals are doing?

TAYLOR: There was a remark made a moment ago about the model not being concerned with social behaviour. I would like to make it quite clear that it does in fact include social behaviour if you think in terms of population dynamics being concerned, not with the actual motivation of behaviour as usually described by behaviourists, but with its outcome in terms of physical movement on a grid—if you think of populations being chessmen moving about on a chessboard, the motives and the physiological mechanisms are not necessary to describe the outcome of the game. All you need to describe is where the men actually moved to. If you have a function which describes the chances of them moving one way or the other on future occasions, or not moving at all, then that may be sufficient for projection. However, you need to know the kinds of behaviour available to each piece—the individual range of variability—and the whole thing doesn't become believable unless you also know something about, not only the behaviour, but the physiological mechanisms behind it.

MARTIN: In conclusion, perhaps I could first say something relevant to the last question. I think that most of us who are concerned about the problem of how these mechanisms operate in terms of the individual, basically return to the question of how such mechanisms could have evolved. However, until we can explain what physiologically makes an individual stop fighting, decide either to accept a subordinate role *in situ* or to move out, we cannot possibly understand how the mechanism evolved. Much of the controversy in this field revolves around discussion of genetical models for the evolution of mechanisms of population control or dispersal. I believe, therefore, that despite your scepticism, one has to work at the individual as well as at the population level and to do this we need people in a large number of different disciplines. All of the papers this morning have been relevant to the problem of how at the individual level these mechanisms operate.

There are just two other comments I would like to make. One is about stress: although we are still at a stage where we don't have much evidence of how stress might operate under natural conditions there is certainly a strong suspicion attaching to stress as a phenomenon. It is worth considering that we have a very strong uncertainty principle operating in that anything we do in order to manipulate an animal's behaviour is also likely to stress it. This applies in the field, where we are trapping, or perhaps even only observing the animals. I would also like to say a little more about migration. I have with Dr Simon Bearder just completed a radio tracking study of bush babies (*Galago senegalensis*) in South Africa. A population of approximately 40 marked known individuals has been followed for 2 years, and we have evidence of males moving out of the area and moving into it. We do not know what provokes this, only that it happens, and an animal which normally lives in a home range of 300 metres wide will simply get up and move a kilometre or two kilometres and then set up another home range elsewhere. In our study area (which has a fairly dense bush baby population) animals also move in and set up home ranges. So there is an attraction of other bush babies as well as a dispersal effect. Once, in Madagascar, when I was working on mouse lemurs I came across a male sportive lemur (*Lepilemura mustelinus*) in the middle of a vast sisal field, which had obviously decided to move. It was halfway across a field at least 2 km wide, but as far as I could see there was no suitable habitat in the direction in which it was heading; the animal had simply taken off. I am convinced that migration is an important part of the picture, though not the only part.

Session Four

Chairman: G. E. FOGG

Discussion on papers by J. B. Birdsell and V. Geist

WYNNE-EDWARDS: Could you say, Dr Birdsell, to what extent the Aborigines were rational or knew what they were doing in population regulation, or to what extent it is all obscured by witchcraft and mumbo-jumbo?

BIRDSELL: I think at this late date, with no new information to come in, and very bad observers to begin with, it is impossible to state. My own feeling is from Dr Greenway's information at Groote Eylandt, that the old granny woman was acting as a social surrogate for her group. In short, this was planned population control. But there is no explicit statement of this—I think one has to look at the consequences and say that a random chaotic system could not produce these uniformities. I think they knew. In short I think there was a very intelligently operated and complex living system.

CLARKE: Could I ask for clarification about the 3 or 4 year spacing of children? Do I understand that it was effected entirely by infanticide, or did they have other methods of contraception?

BIRDSELL: I think the net regulation was by infanticide; there are a few accounts in the early literature of abortions, by pummelling the pregnant abdomen, but no eye-witness accounts, and it has the aura of invention about it, but it has to be granted as a possible alternative. I know of no herbal remedies used. There is one suggestion that wives abstained from sexual intercourse for a period after pregnancy, but certainly the men didn't, and this involves the exposure of other women to pregnancy. The period of weaning is pretty well attested; E. M. Curr, about 1880, sent some questions up country and he got about 150 responses back. The average suggested was that 3–4 years was spent in nursing before actual weaning occurred. I know of no really authoritative statements in the literature of the extent to which suckling suppresses ovulation, but I do recall one horrid account of an Englishwoman who gave birth one month and became pregnant the next. And so I presume that the hormonal control of ovulation in man has broken down to some degree. I would be interested in comments from those of you who do know.

CLARKE: On that last point I think that there are enormous differences between species in the effect of lactation on ovulation. In women it is a notoriously bad contraceptive.

WYNNE-EDWARDS: I believe that some Aborigines practiced sub-incision as a form of contraception. Sub-incision means making a hole through from the outside at the base of the penis into the urethra. The question is: does that interfere with subsequent fertility?

BIRDSELL: There is good literature on this: Victorian Englishmen felt that this 'terrible rite' must have decimated the natives, but the trouble is that tribal populations were the same everywhere whether they sub-incised or not, and indeed the surgery ranged from a hole to a lengthwise split from one end to the other. Apparently this had no effect of diminution on reproduction.

WYNNE-EDWARDS: Why did they do it?

BIRDSELL: The whole question of initiation rites is open to explanations and hypotheses. The pattern of spread, namely origination in the centre and movement towards the peripheries, suggests that people simply became more complex in their mutilations. 'If they do it next door, we will get to do it in time—it may be a generation or two before we get the rights to the songs, but it will come . . .' Particularly with sub-incision, the young men ran away from it—there was by no means easy acceptance of this mutilation. There has been a great deal of Freudian material written about this, but I am not prepared to quote it—I think aboriginally it was just cussed old men inventing worse things to do to young men. I use the term advisedly because T. G. H. Strehlow in his great book on the Aranda songs and legends quotes texts which indicate that the old men had a kind of vengeful feeling as they initiated the young, and that it wasn't all for the social good. There was a certain amount of sadism in this and Strehlow is the one man in the world who would know this. He grew up as an Aranda tribesman, so to speak.

SAXON: Dr Geist, if health is defined in terms of phenotypic development, then one must accept that health (and therefore social mechanisms for the control of populations) must vary between populations. In short, Canada and China (in so far as they vary genotypically) may have educational and social policies for the promotion of health which are quite different.

GEIST: You are right in one important thing, namely that in human beings different social systems probably produce different phenotypes. The ultimate conclusion is that if we want to control health we will have to control the system that generates phenotypes. But I would like to put forward a word of caution. The reason for developing a theory of health is to provide an ideal. If you know, as a professional, that you are going to deprive somebody of health, ontogenetic development, you can sound the warning and ask the question whether you wish to go in that direction. It is not up to me, however, to say what that social direction shall be. I can only state the ideal, namely health; the decision is up to others.

COHEN: I am rather unhappy about Dr Geist's picture of the ice-age man in his perfection, rushing about and capturing enormous bison and so on. Although I believed this picture when I read H. G. Well's *Outline of History*, over the past few years I have come to have quite a different, and I think equally convincing picture, perhaps moderated a little by the belief that he might at some time have been on the seashore. There must have been a lot of women around being selected for apparently quite different qualities. Perhaps they selected the men who could catch them the biggest bison, but most of the time I suspect they were digging up roots and attending to the actual business of keeping the family alive.

GEIST: In the northern environments you must depend for eight months of the year on the resources you can obtain in summer, for root digging and plant collecting are not possible in winter. I know of no primitive herbivorous culture in cold climates which is based on female labour. You have evoked the conventional tribal picture. Many people have proposed that you can learn about the 'natural' lifestyle of man by looking at aboriginal people living today. But which aborigines? Are we to learn from the Caribou Eskimos who every so often starve to death in Northern Canada? Or from the Aleuts of coastal Alaska, many of whom reached ages of more than ninety years, who had a system of medicine and were such superb anatomists that they had a native name for not only every bone but every foramen? We could look at Australian Aborigines or the Ik or the Pygmies. Which group are you going to choose to provide a standard? Professor Bourlière of Paris reported that some native people in equatorial Africa carried something like 3–4 per cent of their body weight of parasites. A very interesting health picture indeed! It is these thought-provoking facts which led me towards the proposals that I am making today.

COHEN: So we have a number of alternatives which all have their attractions. There are people doing all kinds of different things and trying all kinds of lifestyles; undergoing glaciations in the Northern Hemisphere and growing roots in Africa.

GEIST: But do you approve of the Upper Palaeolithic model I was talking about, or do you prefer western occidental culture?

COHEN: I think I like both. I don't know why I have to decide between one or the other.

TAPPER: I was interested to read in your paper that you maintain that these Palaeolithic people probably showed little territorial behaviour. You paint a picture of them living a lifestyle similar to the North American plains Indians. Apart from the fact that they did not possess horses, they were also nomadic and fed on large herbivorous mammals. However, plains Indians showed extremely violent territorial behaviour, and I wondered why you maintained that the Palaeolithic people did not.

GEIST: The American Indian in his exploitation of the buffalo was exceedingly vulnerable to the actions of his neighbours. The neighbour could start fires to deflect the buffalo herd, which means that the tribe not powerful enough to influence his neighbour's behaviour could starve. Whenever you find people highly vulnerable there are incentives for warfare.

CLUTTON-BROCK: Another distinction between man and his earlier ancestors and other mammals is his ability to associate with other species of animal. Man as an ice-age mammal had a social structure that was very similar to that of the wolf. I would like to suggest that these two hunters did collaborate as predators in the Upper Palaeolithic.

GEIST: I haven't involved myself in the question of how wolves could possibly have associated with human beings and what led up to the systematic attempts by human beings to take advantage of them, but that obviously is what must have occurred.

SWANSON: I would like to ask Dr Birdsell whether, in the harsh environment described, infanticide was not an extremely energy-wasteful method of controlling reproduction. Why did these people not evolve some methods of contraception?

BIRDSELL: I am sure infanticide was wasteful. It may have been less wasteful to adult human life, however, than badly conducted abortions or brutally pummelled abdomens. But, in any case, there were no technical facilities for less wasteful methods.

EBLING: Were Aborigines absolutely clear about the relationship between sexual intercourse and reproduction, which would be prerequisite for the invention of contraception?

BIRDSELL: You can get no straight answer, but I'll give you an ambiguous one favoured by those who have known them a long time. The conflict really arises from poor questioning, because where you get no known connection with sexual intercourse, it does seem that the main informants are talking about spiritual values and totemic associations, and if you ask them on the side 'You mean you really don't know . . . ?' you get: 'Oh, yes, but that's old stuff, isn't it?' In short, I think that they dichotomise sexual relations in terms of spiritual and corporal realities.

DEAG: I would like to ask you about infanticide, Dr Birdsell. Am I correct in believing that the graph that you showed us referred to a population under the most inhospitable conditions?

BIRDSELL: They were western desert Aborigines, all living in less than 8 inches rainfall per annum sandhill country.

DEAG: In that case, can you give us any equivalent data for other habitats, for instance near the coast, where the conditions may have been a lot easier?

BIRDSELL: Curr's reports, which covered all sorts of environments, almost universally reported infanticide. It seems to be density-independent and merely a funnelling device to control population recruitment.

CHERFAS: The !Kung bushmen have a birth spacing of about 4 years, which recent studies have shown to be practically optimal. How do they achieve this? Do they practise infanticide?

BIRDSELL: The !Kung bushmen are recorded at various times and by various people. The early accounts describing infanticide differ markedly from those of the last few years. Lorna Marshall, who is a first rate ethnographic observer, stands in between. It has been suggested that they were so gentle and pacific a people that they practised abstinence.

HARCOURT: Professor Geist, your environmental conditions for the healthy man seem almost impossible for most of present-day man to attain. Are you essentially saying that it is impossible for present-day man to be healthy? Are you presenting a hopeless case for the politician or the sociologist to make a plan for a healthy modern-day man?

GEIST: I don't think it's hopeless, but I don't think it's exactly encouraging either.

BLAKE: I have a question for Dr Birdsell. You talked about defending territory. Since warfare and aggression are serious problems for our society, in this well-balanced pre-contact society of the Australian Aborigines, how much aggression was involved in the defence of territory? If not, how did they do it?

BIRDSELL: Defence varies considerably. There is an interesting recent article by Nicholas Peterson who describes the conventional greeting ceremony which did away with a lot of aggressive behaviour. A stranger or group of strangers coming into a settled camp would walk in visibly, sit down a few hundred yards away, and perhaps wait 2 or 3 hours until some old man came out with a container of water to talk and ask what their intentions were. If you came in unannounced and were not known, you were fair game to be speared. They did not patrol boundaries very aggressively, but when found, intruders were very quickly taken care of. And then there is the business of 'is the man known?', 'can he speak our language?', 'do we know what his relationships are in the kinship network?'. The natives themselves were very fearful of intruding into other people's territory uninvited. Invitations were rather freely extended to people who were known, but uninvited people were in great jeopardy.

BLAKE: Did they mark the territories?

BIRDSELL: In the richer countries the territory was marked;

it was known, each man had his boundaries in mind, and presumably from childhood onwards you could recite them. It was a well-defined thing where the country was valuable enough to worry that much about it.

SAXON: May I take the liberty of answering one question which came up because it poses a question back again? I have been working in Tierra del Fuego. The indigenous population there was extremely healthy and extremely territorial before contact with Western man. The question that this raises is whether we see, in other species, conflicts arising where health and territory collapse when two populations of the same species enter the same territory? If Dr Geist wants to restrict himself to human beings, what is the policy for having healthy French, English, and Eskimo in Canada?

GEIST: In my paper I only hinted about the Upper Palaeolithic lifestyle. I did not go into details about what has to be done, or what are the criteria to maximize health. These criteria would include such things as the maintenance of the family structure, and—something less tangible—the ability of individuals to believe and have confidence in what they are doing. The history of occidental cultures, with few exceptions, has been one of brutal destructions of native cultures. It is the period between collapse and reconstruction that witnesses the greatest frequency of diseases, homicides, social problems, and everything else negative. One of the prerequisites for a course of health is the maintenance of a system of accepted beliefs, and values; such systems do not fare very well when several societies collide.

CHARLES KREBS: I would like to raise a general question which has bothered me through most of this Symposium. We have talked about the theme of the symposium being population control in relation to social behaviour, but this afternoon we have had statements made about population control in human beings without any evidence whatsoever that this has, in fact, occurred. We are told that Palaeolithic man controlled his population through these various mechanisms of social behaviour; that the Australian Aborigines controlled their populations by various mechanisms involving infanticide. We have been given no evidence that these mechanisms, which do in fact reduce the *rate* of increase, have any effect in *controlling* population size. The lack of evidence would shock any population biologist who works on other animals or on plants. We have for a number of animal species evidence for self-regulation of populations and for the existence of social mechanisms. The red grouse probably provides the best example at the moment, but there are data of a similar kind for the great tit and for a few species of small mammal. But the existence of a bivariate correlation of 0.87 relating aborigines to their land is not valid evidence that their populations are limited by social behaviour. If that is all we mean by social behaviour limiting population, then I think we can demonstrate this for a host of things. I would like, therefore, to raise the general question, which so far we seem to have avoided: what is the evidence that any of these human groups does in fact control its population by social processes?

GEIST: I agree with you. Examples of social mechanisms that limit populations are virtually limited to the ones you have stated. They are the products of years of very elegant work that most of us can envy and I have indicated earlier that as far as I was concerned, the hypothesis that social mechanisms regulate populations, for instance in ungulates, is at the present still a hypothesis, albeit a good one, on which I am still working.

When it comes to human beings, the ability to experiment is out of the question. We have only weak data based on fossilized specimens, but it does suggest that human beings, for a pretty long time, somehow escaped phenotype deterioration. Granted that the phenotype deteriorates greatly if adequate nutrients during the pre-and post-natal growth are not available, the conclusion is inescapable that something intervened, so that Palaeolithic people maintained excellent phenotypic development over a very long time span. That something can only be a social mechanism. But did they control the population, in the sense of governing numbers accurately? I do not know. I doubt it, for the reason that the Upper Palaeolithic was also the time of maximum human dispersal. There was reproductive excess and human beings spread and dispersed; the population was not necessarily static. At the same time I have not the slightest doubt that in harsh environments, and we have some evidence from Eskimos about this, whether a child ought to be allowed to live, or killed for the benefit of all, was carefully evaluated.

CLARKE: I want to come back to Charles Krebs' provocative remarks. What is sufficient evidence to establish the social regulation of population size for any species, and how much of that evidence could be called direct, and how much inference?

TAYLOR: I also wish to put two questions to Dr Krebs. First, how many generations are required to demonstrate 'stability' and second, are there any animals that live realistic lives in experimental conditions?

CHARLES KREBS: To answer Dr Taylor first, field populations certainly do not exist in experimental laboratory conditions, and I think experimental situations are useful only in so far as they mimic some essential quality of the field situation. This is illustrated, for example by the work on small rodents, in which we found enormous densities in laboratory populations but not in the field, and an enormous difference between our confined populations in the lack of dispersal. This points to *dispersal* as a population process which might explain some of the differences between the laboratory and field situations.

On the question from Dr Clarke of what kind of evidence we need—I think the best answer has been given in a review by Watson and Moss (1970)*. They suggest that you demonstrate that there are surplus animals in a population, excluded by the social system, that these animals can breed when you provide space for them, and that the population can respond by these various mechanisms of social behaviour to changes in other variables. They note the incredible difficulty of getting these data, which, as I said, have been provided for the red grouse and the great tit and by ourselves for several species of small rodents. I think the emphasis has to be laid on experimental procedures for demonstrating surplus individuals in populations.

BIRDSELL: The thrust of comments by Drs Cohen, Krebs, and Clarke is that the Australian Aborigines belong with all the other vertebrate populations and are to be judged by the same criteria. They expect human hunters to show a category of surplus individuals, presumably males, who are systematically excluded from normal reproductive activities through the opera-

* Watson, A. and Moss, R. (1970) Dominance, spacing behaviour and aggression in relation to population limitation in vertebrates. In *Animal Populations in Relation to their Food Resources*, ed. Watson, A. pp. 167–218. Oxford: Blackwell Scientific.

tion of the social organization of the population. Although these expendable persons could under other circumstances breed if allowed, it is their fate to disperse and suffer heavy losses through predation. They consider this the primary way by which population homeostasis is to be achieved.

Over the years I have read just enough studies in British population ecology to have developed a profound respect for the ingenuity of its methods and the beauty of its results. But the dogma of what in a jocular vein might be referred to as the 'mouse-grouse' model is not applicable to human hunters. In all logic I must reject it for them.

Human hunters differ in a number of polar ways from the other vertebrate populations studied by ecologists. With the evolution of symbolic speech *Homo sapiens* entered a unique eco niche, and developed new types of adaptive behaviour and social organization. Man is a long-lived species whose generations span twenty-five or more years, instead of one to several. He is subject to little or no interspecific predation. Population numbers among hunters are consciously controlled through infanticide, usually preferentially female, so that demographic stability is achieved with no need for dispersion.

Ethics alone suffice to prevent experimental methods being practiced among human populations. Further, no generalized hunters function any longer in undisturbed scenes in which the tensions of the original ecological system remain intact. As a consequence, whereas time-calculus is the appropriate dimensional framework for the study of short-lived vertebrates, with man research must be conducted in space-calculus. At best human memory recall and speech together provide data on no more than one or two past generations.

This array of critical differences places human hunters outside of the logic of the classic 'mouse-grouse' population ecology. I have analysed the living system of the Australian Aborigines in terms of the data available. The best data are both hard and extensive. Supplementary qualitative data tend to confirm the operation of their system as both well-controlled and homeostatic. Several of the discussants seem to have forgotten that David Lack pointed out some years ago that differences between human and other vertebrate populations are greater than their similarities.

CHARLES KREBS: I quite agree. I think that the primatologists have difficulty working on population studies—and students of man are even worse off.

WYNNE-EDWARDS: I wonder, with regard to what Dr Krebs has said, if it is right and fair to exclude all laboratory experiments? If you think of the range of laboratory insects like *Tribolium* or *Drosophila*, you might say that their regulative processes are not what you would call social, but they are certainly self-contained and self-operated. *Drosophila* shows typical aggressive behaviour, as I stated in my paper (page 11), and in the experiment I quoted, more than half the newly emerged adults were apparently excluded from breeding by a dominant minority.

CHARLES KREBS: I do not know whether experimental studies of insect populations in the laboratory have been of much use in understanding the regulation of insect populations. I am not an insect ecologist, but my feeling is that they have not. I do know that work on laboratory populations of rodents has been minimally useful in understanding what rodents are up to in the real world.

SWANSON: I would like to address a question to the primatologists. One of the things that struck me about chimpanzees is that they have very good spacing between births. The babies were born 3–4 years apart, and as far as I could gather the researchers did not seem to understand how the mechanism for spacing operated. It seemed that the females failed to go into oestrus while they had a youngster with them. This seems a very good arrangement.

WRANGHAM: To answer briefly; there is certainly a relationship between the presence of an infant and the probability of oestrous cycles occurring, but, as in man, it is very weak. The time between birth and the resumption of oestrus has varied from 11–81 months, from five years' data for Gombe chimpanzees. But it does seem clear that lactation is the major factor responsible for the cessation of oestrus, just as in man. Studies in the Kalahari and Rwanda, for instance, have indicated that better nourished populations have a more rapid return to oestrus following birth.

FOGG: I think we ought to bring the discussion to a close. When the organizers of this symposium invited me to take the chair at the last session I hope it was understood that no intelligent summing-up could be expected from one who only studies phytoplankton. But I must say that this has been an extraordinarily interesting and thought-stimulating session and so, I am sure, was the whole symposium.

Index

Note: numbers in italics refer to tables and figures.